PROTOGAEA

Translated & Edited by

CLAUDINE COHEN & ANDRE WAKEFIELD

The University of Chicago Press   Chicago & London

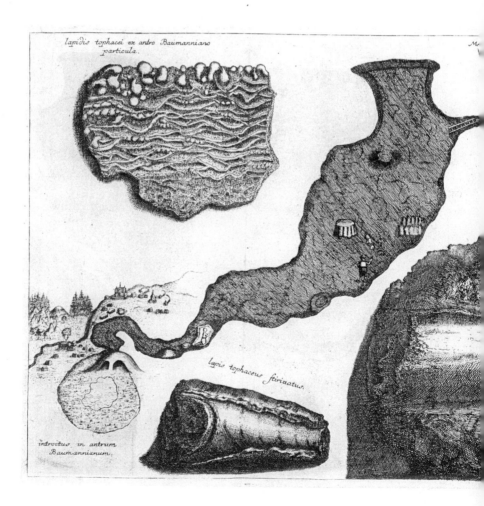

lapidis tophacei ex antro Baumanniano
particula.

lapis tophaceus stiriatus.

introitus in antrum
Baumannianum.

PROTOGAEA

Gottfried Wilhelm Leibniz

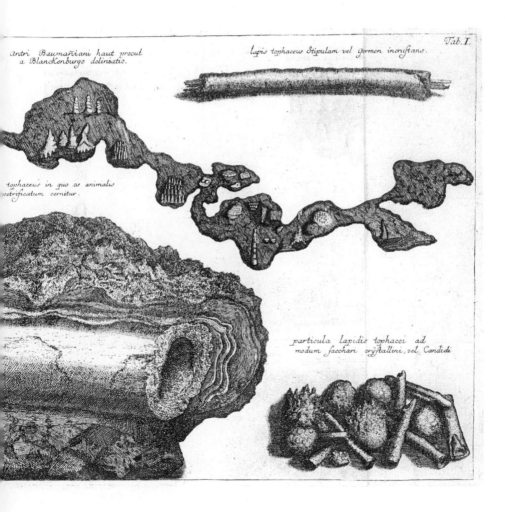

antri Baumanniani haut procul a Blanckenburgo delineatio.

lapis tophaceus Stipulam vel germen incrustans.

tophaceus in quo as animalis petrificatum cernitur.

particula lapidis tophacei ad modum sacchari crystallini, vel Candidi

Tab. I.

GOTTFRIED WILHELM LEIBNIZ
(1646–1716)
Claudine Cohen is professor at the
École des Hautes Études en Sciences
Sociales, Paris, and the author of *The
Fate of the Mammoth: Fossils, Myths,
and History*, also published by the
University of Chicago Press. Andre
Wakefield is assistant professor of
history at Pitzer College.

The University of Chicago Press, Chicago 60637
The University of Chicago Press, Ltd., London
© 2008 by The University of Chicago
All rights reserved. Published 2008
Printed in the United States of America

17   16   15   14   13   12   11   10
    2   3   4   5

ISBN-13: 978-0-226-11296-1 (cloth)
ISBN-10: 0-226-11296-9 (cloth)

Library of Congress
Cataloging-in-Publication Data
Leibniz, Gottfried Wilhelm, Freiherr von,
1646–1716.
    [Protogaea. English & Latin]
    Protogaea / Gottfried Wilhelm Leibniz ;
translated and edited by Claudine Cohen and
Andre Wakefield.
        p. cm.
    Includes bibliographical references and index.
    ISBN-13: 978-0-226-11296-1 (cloth : alk. paper)
    ISBN-10: 0-226-11296-9 (cloth : alk. paper)
    1. Geology—Early Works to 1800.
2. Paleontology—Early works to 1800.
3. Historical geology—Early works to 1800.
4. Earth—History—Early works to 1800.
I. Cohen, Claudine. II. Wakefield, Andre.
III. Title.
    QE25.L513 2008
    551—DC22

                                    2007033903

*For
Zachary,
Xenia,
and Eli*

CONTENTS

ACKNOWLEDGMENTS

This book project benefited from the generosity of many people. Roger Ariew and Daniel Garber offered invaluable advice and criticism, especially during the early stages of research and writing. William Newman and Lawrence Principe graciously offered their much-needed expertise at various points. Martin Gierl provided research support in Germany. We would also like to express our gratitude to others who helped us along the way: Luca Ciancio, Michel Fichant, Ernst Hamm, John Holloran, Peter Gay, David Oldroyd, Kenneth Taylor, and Rhoda Rappaport. In addition, Ryan Boynton, Molly Ierulli, Paul Mueller, William Rodarmor, and Annelies Wouters helped in various ways with the Web site and translation. Finally, our students and colleagues at the École des Hautes Études en Sciences Sociales and the Claremont Colleges have provided valuable encouragement and criticism.

This book was made possible by a collaborative research grant from the National Endowment for the Humanities (RZ-20765-01), and we warmly thank James Herbert for his kind support. The Dibner Institute for the History of Science and Technology in Cambridge, Massachusetts, provided us with research fellowships and hosted several meetings on *Protogaea*. The project received other support from Princeton's Institute for Advanced Study, the Getty Research Institute in Los Angeles, and Pitzer College in Claremont, California.

For access to collections and archives, we would like to thank the Landesbergamt Clausthal-Zellerfeld, the Forschungsbibliothek Gotha, Honnold Library Special Collections, and the Burndy Library. For permission to reproduce plates and images, we thank the Gottfried Wilhelm Leibniz Bibliothek in Hannover and the Huntington Library in Pasadena.

We wish to express our gratitude to the editors at the University of Chicago who have seen this project through to its conclusion. Susan Abrams offered crucial support for the project, Catherine Rice saw it through the middle stages, and Christie Henry has guided things to a successful conclusion.

C.C. and A.W.
November 2007

Protogaea, written between 1691 and 1693, was one of several works from the late seventeenth century to offer a conjectural history of the earth. In it, Leibniz hypothesized about the origins of mountains, volcanoes, and springs. He contemplated the classification of minerals. He demonstrated the organic origin of fossils and tried to explain their presence on mountaintops, and he included a series of engraved plates representing both the fossil remains of animals and cross sections of the caves in which they had been discovered. Leibniz grappled with contemporary works on the earth, including René Descartes's *Principia philosophiae*, Nicolaus Steno's *Prodromus*, Thomas Burnet's *Telluris theoria sacra*, and Agostino Scilla's *La vana speculazione disingannata dal senso*.[1] He quoted at length from Georgius Agricola, Athanasius Kircher, and Johann Joachim Becher, drawing on a range of evidence that relied not only on theories and textual arguments but also on the practices of excavation, collection, reconstruction, authentication, and experiment. Though deeply committed to general laws and causal explanations, *Protogaea* also relied on the careful examination of fossil objects as "documents" for the history of the earth, drawing on the same methods that Leibniz used in his historical works. The work also echoed major themes in Leibnizian philosophy, seeking even in geological catastrophe and cataclysm the order and meaning of that best of possible worlds.

But the sources and context for *Protogaea* spilled out well beyond the confines of philosophy and natural history, into the mines of the Harz and the secret archives of the Hannoverian dukes. Leibniz, that is, worked intensively—obsessively even—to introduce wind power as a driving force in the silver mines of the Harz Mountains. This project, which took him away from Hannover to the towns of the Harz for months at a time, provided privileged access to the minerals and fossil objects of Lower Saxony. Finally, *Protogaea* is imbued by the knowledge, sensibility, and ambition of Leibniz the court historiographer. It was, after all, intended as a preface to his monumental history of the House of Brunswick.

It would be wrong, therefore, to read this text as an abstracted scientific

1. Descartes [1644] 1983; Burnet 1681; Scilla 1670.

or philosophical treatise concerned only with universal questions. In fact, the text is thoroughly embedded in a series of local contexts, from the mining towns of the Harz to the wells of Modena, from the Weser River near Minden to the caves of Quedlinburg. And yet the universal is always lurking just behind these particulars. The virtue and the promise of the work lie largely in the way Leibniz weaves these worlds together. That balance between particular and universal, between local and global, is present from the very first section of *Protogaea:* "When everyone contributes curiosity locally," Leibniz wrote, "it will be easier to recognize universal origins."

Leibniz in the Harz

"We occupy the highest region of lower Germany, one that is especially rich in metals."[2] These words, from the first section of *Protogaea,* suggest the ambitions that first drew Leibniz to the Harz silver mines in 1679. For some six years, between 1680 and 1686, he worked tirelessly to install his inventions there, but his efforts ultimately failed. Though Leibniz invested countless hours and much of his own money in these projects, Duke Ernst August finally called the whole thing off in 1686. Leibniz never gave up on the Harz, however, and he returned there in 1693 to begin a new series of experiments. *Protogaea,* written sometime between his two projects in the Harz, reflects Leibniz's detailed knowledge of the region and its mines. In it, he described the formation of the globe in terms of mining and smelting operations, and he drew on detailed local knowledge of the Harz to fashion a plausible history of the earth.

In recent years, scholars have emphasized the significance of mines and mining for *Protogaea,* but the detailed history of Leibniz's ambitions in the Harz has received less attention.[3] Between 1680 and 1686, Leibniz visited the Harz Mountains more than thirty times and spent almost three full years there.[4] In the late seventeenth century, the journey from Hannover to the mountains was arduous, and Leibniz made it repeatedly, suffering through the heat of the summer and cold of the winter to

2. *Protogaea,* §I.

3. See Elster 1975; Scheel 1991; Cohen 1996; and Hamm 1997. The German Academy edition of Leibniz's writings, which treats the Harz material as a special case, reproduces only a fraction of available archival sources on Leibniz's activities there. See Ritter 1938, xxviii–xxix.

4. Ritter estimates that Leibniz spent a total of 165 weeks in the Harz between 1680 and 1686. See Ritter 1938, xxvii; Aiton 1985, 108.

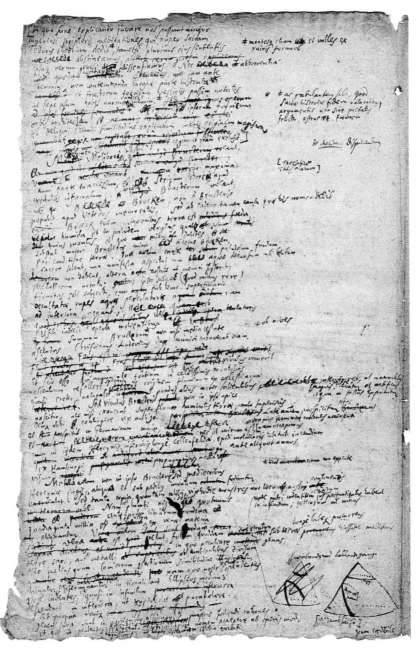

FIGURE 1

Page from the "A manuscript" of *Protogaea*, in Leibniz's hand, with original sketches. See *Protogaea*, §VIII. (Gottfried Wilhelm Leibniz Bibliothek, Hannover.)

spend time in the mining towns there. These were productive years for Leibniz: he published papers in the *Acta eruditorum*, developed the differential calculus, created diplomatic strategy for the House of Hannover, worked on reconciliation between the Catholic and Protestant churches, and composed the *Discourse on Metaphysics*. But none of these activities demanded as much time and energy as his efforts to harness wind power for the Harz silver mines.

It is difficult to overstate the importance of Harz silver to the dukes of Hannover and Brunswick. Long the backbone of the sovereign's finances, the Harz mines had fallen on hard times during the Thirty Years' War, when many productive veins had to be abandoned. By the 1670s and 1680s, reinvestment in the mines had started to reverse that trend, and ambitious entrepreneurs imagined a return to the legendary silver yields of the sixteenth century. For Leibniz, the Harz presented a perfect testing ground for his inventions, and in 1679 he secured Duke Johann Friedrich's support for a new windmill technology that would supplement waterpower in the mines.[5] During wet seasons the water supply was more than adequate to power the wheels and pumps that kept things running, but during dry seasons work could stall entirely. In order to mitigate the problems caused by chronic water shortages, successive governments in Hannover and Brunswick had financed and constructed a complex system of holding ponds, which collected rainwater at high elevation for use during dry spells. Still, water shortages continued to plague the Harz mines well into the nineteenth century. If Leibniz could develop a reliable wind-based technology in the Harz—and he promised that he could—it would add tens of thousands of taler to the treasuries of Hannover and Brunswick.

The first mention of Leibniz's mining venture appeared in a 1678 memorandum to Duke Johann Friedrich.[6] Mentioning few specifics, Leibniz suggested that he had a plan to increase dramatically the output from the Harz mines and explained that he wished to visit them. By 1679, Leibniz had clarified the plan in writing. He hoped that, thanks to his inventions, the pumps would remain in constant operation and mining could continue during all seasons.[7] Leibniz came from a family with old ties to the

5. Ritter, 1938, xxxiv.
6. Leibniz [1678] 1923–, ser. 1, vol. 2, 78–89.
7. Ibid., 53–54, 79–89, and 195–196; Aiton 1985, 87–88.

mines.[8] Thus, though he may have been an outsider to the Harz, his family connections offered access to the peculiar culture of central Europe's silver mines.

A trip to Holland in 1676 got Leibniz thinking about wind power, for it was there that he saw the windmills used to drain flooded lands.[9] Windmills had long been used in the mines to pump air into the shafts, so that miners would not asphyxiate. There had been other experiments with wind power in Germany too. In 1617, for example, the famous Clausthal mining official Georg Engelhard von Löhneyß had described a machine driven by the wind, and Leibniz probably knew about it. Other officials, like the mining councillor Peter Hartzingk, who had studied in Leiden, were building models and conducting experiments with wind power during the early 1680s. Leibniz's novel twist involved using wind to reclaim the force of falling water. He hoped, that is, to capture water as it drained from higher elevations; he would then use the force of the wind to pump that water back up to the holding ponds, where it could be stored for future use. In short, Leibniz hoped to improve the entire water management system in the Harz by supplementing it with his windmills.[10]

Leibniz arrived in Clausthal, administrative center of the Hannoverian mining administration, in 1680 to begin his experiments with wind power at the Catharina Mine. It was not long before he had angered and alienated many of the local miners and officials there. He complained that they were obstructing his plans, and even sabotaging his work; they complained that his work was impractical and would waste money. By the middle of 1683, the total costs of the project, which were to be shared evenly between the duke, Leibniz, and the mining office, had risen to 2,270 taler—about four times Leibniz's annual salary. Experiments in 1684 proved indecisive or failed, largely because of a lack of wind. Leibniz then developed a new, more sensitive horizontal windmill. During

8. Leibniz's grandfather, Ambrosius Leibniz (1569–1617), had served as a mining scribe (*Bergschreiber*) in Altenburg, his great-grandfather, Christoph Leibniz (1537–1587) had been a mine overseer (*Bergmeister*) in Berggießhübel, and another great-grandfather, Heinrich Dauerlein (d. 1595), had served as a mine accountant (*Bergzehntner*) in Eiland. Even Leibniz's great-great-grandfather, Heinrich the Elder, had served as a Saxon mining official. Cf. Horst and Gottschalk 1973, 36. For an overview of mining terms, and difficulties with translation, see Hamm 2001, 286.

9. Leibniz 1923–, ser. 1, vol. 4, 53, 61.

10. See Horst and Gottschalk 1973, 37–44; Calvör 1763, 101–110.

1684 and 1685, the battle between Leibniz and the Clausthal mining office intensified, as the mining officials there started to demand real results. The wind machines seemed to be improving, but the mining office insisted that Leibniz should now, after five years, either put up or shut up. Finally, in April of 1685, the duke ordered Leibniz to stop further work on his wind machines. It had been a costly and bitter affair for everyone involved, and the duke wanted Leibniz back in Hannover to attend to his official duties as court librarian. Leibniz's biographers have mostly created the impression that a group of small-minded mining officials stubbornly blocked Leibniz's promising innovations. "Though he had been beaten by the nature and obstinacy of men," writes one of them, "his optimism remained undiminished."[11] The archival record of the battles between Leibniz and Clausthal's mining officials, housed today at the *Oberbergamt* in Clausthal and still largely unpublished, presents us with another possibility. Perhaps, as Clausthal's mining officials suggested, Leibniz was an outsider from Hannover seeking to impose his impractical innovations on seasoned experts who knew better.[12]

The frequent references to mines and miners in *Protogaea* make it clear that Leibniz's failed project in the Harz had lasting significance for his "natural geography."[13] Moreover, his metaphysics of force and work takes on new meaning in the context of the Harz, for Leibniz was, after all, concerned to harness the work of water and wind in the most efficient possible way. With their massive waterwheels constructed hundreds of feet below the earth, and their large holding ponds built near the tops of mountains to maximize the force of the water, the mines of the Harz were the big technology of seventeenth-century central Europe. At the time, they represented huge capital investments and were the pride of princes and dukes. Can it be any wonder that an ambitious young savant like Leibniz wanted to be involved? As he wrote Duke Johann Friedrich in 1679, "the Harz is a true source of experiences and discoveries in mechanics and physics; and, my lord, I consider myself able to discover more with five or six men of practice, who might be employed in these regions, than with twenty of the most learned savants in Europe."[14]

11. Aiton 1985, 114. See also Ritter 1938, ser. 1, vol. 3, xxix.

12. Cf. Oberbergamt Clausthal-Zellerfeld, Fach 761, Acta 27 and 35; and Fach 762, Acta 27. In 1991, the Berlin Academy published a comprehensive supplementary volume on Leibniz and the Harz for the years 1692–1696.

13. On Leibniz's definition of "natural geography" see below, "Introduction," xxxiii.

14. Leibniz 1923–, ser. 1, vol. 2, 126–128.

Leibniz's "Theory of the Earth" and Its Metaphysical Implications

The years between 1660 and 1720 witnessed an extraordinary efflorescence of speculations, hypotheses, observations, and debates about the formation of the earth and its material history. Following Descartes's lead in the *Principia philosophiae* (1644), many writers offered accounts of the formation of the earth based on rational, mechanical principles. The development of these systems, more or less rooted in empirical observation, included problems ranging from the origins of mountains and the causes of earthquakes to volcanic eruptions and the motion of the sea. Indeed, such seemingly mundane issues had serious religious and philosophical implications, as did the apparently futile debates over "shells and systems built on shells." [15] At stake were several fundamental questions: the status of the biblical narrative as literal truth, the creation of the earth and its inhabitants, the permanence of this creation, and the place of human beings in the universe. [16]

Descartes, first in *Le Monde* (1633) and then parts 3 and 4 of the *Principia philosophiae* (1644), had explained the formation of the earth in mechanical terms, relying on three elements and a few principles to describe the development of the universe, from initial chaos to the present day. [17] Descartes envisioned the earth as a sun in a universe of vortices, animated by circular motion in a plenum. Over time, he argued, opaque spots spread across the surface of the globe, until a crust formed around it. After that, distinct concentric layers formed, eventually collapsing under their own weight to create oceans and mountains. Descartes's narrative was powerful, and it attracted others who wanted to write "theories of the earth" in both historical and physical terms. English diluvialists like Thomas Bur-

15. This is the title of an article in Voltaire's *Dictionnaire philosophique* (1768), in which he derided these geological speculations.

16. During the past few decades, scholars have produced a rich body of scholarship on these "theories of the earth," tracing them through the eighteenth century and investigating their role in the constitution of geology during the nineteenth century. See Roger 1968; Rudwick 1972; Morello 1979; Rossi 1984; Gould 1987; Ellenberger 1988–; Laudan 1987; and Cohen 2008. Paolo Rossi (1984) linked debates over terrestrial history with those about human history.

17. Though Descartes wrote *Le Monde* in 1633, he withdrew it from publication after hearing about Galileo's condemnation in Rome. It was first published in 1664, after his death. Descartes [1633] 1979. For his more detailed conjectural account of the formation of the earth, see Descartes [1644] 1983, 181–219 (IV).

net took Descartes's conjectural history of the earth and wove it together with the narrative scheme of Genesis.[18] In their hands, the biblical flood became the main cause of terrestrial events like the collapse of the earth's crust and the formation of mountains and oceans.

Protogaea constituted both a continuation and a criticism of Descartes's project. By the end of the 1680s Leibniz had already read and annotated the *Principia philosophiae,* criticizing the foundations of its physics and metaphysics and pointing out specific errors and weaknesses.[19] *Protogaea* thus contained many criticisms of Cartesian physics and cosmology, some explicit and some implicit. Leibniz intended his account of the earth's formation to be both logical and historical, based upon observable evidence and particular detail.

Like Descartes's *Principia,* the early sections of *Protogaea* reconstructed the formation of the earth from its beginnings. The globe had originally been a molten mass, becoming harder as it cooled. "Vitreous" materials— rocks and sands, the "great bones of the earth"—resulted from this initial fusion. The cooling process produced enormous "blisters," filled with air or water, which, upon collapsing, yielded mountains and valleys. As the earth cooled further, "aqueous vapors" condensed into water, which combined with the salts to form the seas. The weight of the waters then caused sections of the earth's crust to collapse, resulting in great floods. These floods, in turn, left behind sediments that hardened as the waters subsided. Earthquakes, smaller floods, and volcanic eruptions followed in the wake of these great events. Finally, this violent early history gave way to a more peaceful epoch, suitable for humanity. Unlike Descartes, however, Leibniz provided detailed descriptions of the physical and chemical processes—flooding, combustion, evaporation, condensation— involved in such transformations.

Other parts of *Protogaea* introduce a different narrative, one in which Leibniz focuses on the history inscribed in different layers of the earth. While telling of the well diggers of Amsterdam or Modena, for example, Leibniz follows a continuous succession of terrestrial layers from present to deep past, sketching what we would today call "stratigraphy." It was possible, through this method, to reconstruct the sequence of events that occurred as the earth formed. Leibniz's approach transformed the differ-

18. Burnet 1681.

19. "Animadversiones in partem generalem Principiorum Cartesianorum," in Leibniz 1960, vol. 4, 350–392.

ent episodes represented in this spatial succession into a temporal and causal narrative. "In all likelihood, there was once a seafloor where shells now lie, at a depth of more than one hundred feet. Repeated floods and catastrophes have thrown all the layers of clay and sand upon this floor, while the deposits of earth arose during the intervening periods. The sea, driven back, retreated for a time. But ultimately, insisting on its right, the sea once again burst the dams, flooding the lands and flattening the forests, whose ruins are now revealed during excavations. For us, therefore, nature stands in place of history."[20] Here an archaeological and historical perspective allowed Leibniz to reconstruct the events of the past by reading the earth through the excavations of the present. Using knowledge gleaned from his experiences in the Harz mines, he looked for evidence in the folds and layers of the earth, in the disposition of mineral ores, and in marine fossils. Moreover, he proposed to verify his claims about the formation of the earth, of minerals, and of fossil objects through chemical experiments.

Leibniz was troubled by the relationship between Descartes's account and the biblical narrative of creation. Descartes presented his history of the earth as a fiction or fable, explaining that the causes he invoked were at once true and false: true with regard to mechanical laws and false with regard to revelation.[21] Leibniz, who considered such an attitude theologically dangerous and methodologically flawed, presented his account not as a fiction but as the framework for a real history. He turned to the Bible for support, finding justification there for his notion that the earth had formed through the alternating actions of fire and water. Leibniz also stressed the similarity between his theses and those of Burnet and Steno, insisting that their writings contributed "through natural arguments, and not without benefit for piety, to a belief in the sacred history and the universal flood."[22] But his account also differed from those of the diluvialists. While they regarded the biblical flood as the pivotal event in terrestrial history, Leibniz wrote of *several* great floods in the earth's history, both before and after the appearance of man. These were not miracles, supernatural punishments inflicted by God upon humanity. Rather, they appear in *Protogaea* as mechanical phenomena.

20. *Protogaea*, §XLVIII.
21. Descartes [1644] 1983, 84–85 (III). On the problems raised by the Cartesian fiction, see Grene 1985; and Cavaillé 1991.
22. *Protogaea*, §VI.

If Leibniz shared the Cartesian vision of an earth in ruins, he rejected any suggestion of primordial chaos, for that threatened to eliminate God's role in creation.[23] "God makes nothing without order," he wrote, revisiting a central theme from the *Discourse on Metaphysics* (1686).[24] It was the starting point of *Protogaea*, since history revealed the unfolding of God's plan in time. As Leibniz explained to Louis Bourguet in 1714, limited observers of this process could not comprehend its global necessity:

> When I affirm that there is no chaos, I do not at all intend to say that our globe or other bodies have never been in a state of external confusion: for this would be contradicted by experience. The mass Vesuvius expels, for example, is such a chaos; but I mean that anyone who had sense organs penetrating enough to perceive the smallest part of things would find everything organized. . . . For it is impossible that one creature is capable of penetrating everything simultaneously in the smallest parcel of matter, since the subdivision goes on infinitely. In this way, the apparent chaos is but a sort of distance: as in a pond full of fish, or better, as in an army viewed from afar, where it is impossible to distinguish the order observed there.[25]

In Leibniz's view, God had quite literally infused the world with force, itself the proximate cause of appearances in the physical world.[26] *Protogaea* reveals a living earth, bristling with potential energy, and then, later, a ruined earth whose history is written in its fossils and strata. The young, double-vaulted earth, filled with layers of water and air, contained within itself, from that first day, all the forces that would ruin and transform it. For Leibniz, force was the protagonist of terrestrial history.

Leibniz often rejected the notion of dead matter, collapsing the distinction between animate and inanimate worlds.[27] As he wrote in the *Monadology* (1714), "there is nothing fallow, sterile, or dead in the universe, no chaos and no confusion except in appearance, almost as it looks in a pond

---

23. See letter to Louis Bourguet of 22 March 1714 in Leibniz 1960, vol. 3, 565–566. Cf. Paolo Rossi's comment in Rossi 1984, 55.

24. Cf. Leibniz 1989, 39–40 (§6–7).

25. Leibniz 1960, vol. 3, 565–566.

26. On force in Leibniz, see Garber 1995, 281–298.

27. The status of matter, and its relationship to body, substance, and monads, is the subject of considerable debate. See Rutherford 1995, 143–153.

at a distance, where we might see the confused and, so to speak, teeming motion of the fish in the pond, without discerning the fish themselves." [28] In *Protogaea*, Leibniz sometimes treated the earth itself as a kind of living organism, though it is unclear how literally he meant it.

> Not many years ago a certain soldier and traveler appeared, who called himself Alexander Achilles. He dared to draw the bones of the great mother and the veins running through her body, as if he had inspected the naked earth and, as the only one among mortals, had scrutinized "all streams beneath the mighty earth that glide." [29] . . . There is no doubt that something like the formation of plant or animal occurred when the creator wove the first fabric of the tender earth. But this got so confused and distorted by fires and floods and collapses in the earth's surface, which is like a skin, that it became very difficult to recognize. [30]

But if the earth could be compared to a plant or an animal, it was also filled with ruins, which one could read like artifacts or archives to recreate its violent history, replete with catastrophic collapses, floods, and rock slides. Here, in the collapse of massive rock domes and cataclysmic floods, was a conjectural history of force infused with Leibnizian dynamics.

Important philosophical issues were at stake here, issues that bore directly on Leibniz's metaphysical system. Like preformed living beings, which carried in embryo all the marks of their future development, the world too had a physical destiny inscribed in its beginnings; its history represented the disclosure of God's plan. [31] In his *Theodicy* (1710), Leibniz used a mathematical image to stress that the history of the world could reveal only the gradual unfolding of God's plan, and that a finite, partial observer could not hope to understand the global necessity of that divine plan. "It should be no cause for astonishment," he wrote, "that I endeavor to elucidate these things by comparison taken from pure mathematics, where everything proceeds in order, and where it is possible to fathom

---

28. Leibniz 1989, 222 (§69).
29. Virgil, *Georgics*, 4.366.
30. *Protogaea*, §VIII.
31. See the letter to Louis Bourguet of 22 March 1714 (Leibniz 1960, vol. 3, 565). See also §13 of the *Discourse on Metaphysics* (Leibniz 1989, 44–46).

them by a close contemplation which grants us an enjoyment, so to speak, of the vision of the ideas of God"[32] The universe, Leibniz suggested, was like a series of numbers: it might appear irregular and unlawful even as it conformed to a definite formula. This mathematical image served as the prelude to a brief summary of Leibniz's theory of the earth, as he briefly sketched the origins of the planet: the fiery globe and its burnt crust, the great rains that washed over the cinders of the earth, and the salty oceans that then gathered in the cavities of the earth.[33] Moreover, in *Theodicy* he explicitly linked the themes of *Protogaea* to his ruminations on order, disorder, and possible worlds.

> Sundry deluges and inundations have left deposits, whereof traces and remains are found which show that the sea was in places that today are most remote from it. But these upheavals ceased at last, and the globe assumed the shape that we see. Moses hints at these changes in few words: the separation of light from darkness indicates the melting caused by the fire; and the separation of the moist from the dry marks the effects of inundations. But who does not see that these disorders have served to bring things to the point where they now are, that we owe to them our riches and our comforts, and that through their agency this globe became fit for cultivation by us. These disorders passed into order.[34]

The lesson was clear: destructive episodes and catastrophes had meaning from a universal perspective. Floods and earthquakes had their purpose in the eyes of God, and history itself, with all its monstrosities and disasters, might in principle be reduced to a mathematical formula that reflected the divine plan.

After the horrors of the Lisbon earthquake in 1755, Leibniz's so-called optimism would be ridiculed by Voltaire in *Candide*. And yet, for Leibniz, it was really a matter of perspective. In other words, terrestrial history could be written either as a succession of contingencies or as the product of natural laws at work. Years later, Immanuel Kant would emphasize this distinction between a *description of nature* from the present (based on the human view) and a *history of nature* from its origin (based on the divine view).[35] Both perspectives appear in *Protogaea*, with the deductive

32. Leibniz [1710] 1966, 124 (§241).
33. Ibid., 125–126 (§244–245).
34. Ibid., 126 (§245).
35. Kant [1786] 2004, 3–6.

narrative from general causes and natural laws taking precedence in the first chapters, and the inductive historical approach dominating the later chapters.

The Nature of Fossils

In his discussion on the nature of fossils and the formation of the earth, Leibniz was deeply influenced by the works of the Danish anatomist and naturalist Nicolaus Steno, whom he read, and probably met, as early as 1678.[36] Steno, a follower of Descartes, studied anatomy in Amsterdam and eventually moved to Florence, where he served as physician to Grand Duke Ferdinand II. While in Florence, Steno traveled widely in Tuscany, analyzing the region's geological features. When he dissected the head of a shark in 1666, Steno demonstrated that glossopetrae, traditionally regarded as petrified snakes' tongues, were in fact sharks' teeth. How could they have been inserted into layers of the earth, often far from the sea? In his *Prodromus*, Steno concluded from "the enclosure of a solid inside of another solid, [that] they had been formed, not simultaneously, but successively."[37] In other words, fossil shells and fossil fish had existed *as such* before being embedded in layers of the earth. As a result, he concluded that these objects had not been, as many believed, formed inside the earth. This spatial relationship of enclosure thus expressed a temporal relationship of succession, so that time became implicated in the formation of fossil objects. This simple principle also suggested that one could explain the formation of the earth's layers on the basis of their spatial disposition, and that one could use the fossil remains embedded in these layers as clues to the past.

In *Protogaea*, Leibniz adopted Steno's method. Just as Steno had studied the formation of Tuscany, Leibniz now used the "natural geography" of Lower Saxony to construct a general theory of the formation of the globe. "Certainly, if the earth was liquid at the beginning, it would have been uniform; and it agrees with the general laws of bodies that solids harden out of liquids, as witnessed by solids enclosed in solids, and by layers and kernels enclosed in the fissures of the earth, like veins in rocks and gems in stones. And then there are the scattered vestiges of old things, of plants, animals, and artifacts wrapped in a new coat of stone. Thus, what we now perceive as hard appeared only later and must certainly

36. See Roger 1968, 137–140.
37. Steno 1669.

have been liquid once."[38] Here Leibniz quoted Steno's title directly, while reiterating the principle articulated in *Prodromus*. The presence of solid bodies—minerals, gems, plants, and animals—inside other solid bodies implied the importance of time as an operative force. Fossil objects that resembled living beings, therefore, arose neither from chance nor from "games of nature," nor even from some mechanical process that shaped them under the earth. Rather, their existence and position implied successive episodes of burial and sedimentation.

Leibniz stressed the importance of distinguishing among different kinds of objects found in the earth. On the one hand, there were inanimate bodies, such as crystals and "polygonal shapes," the formation of which could be explained by "external contiguity"; on the other hand, he strongly criticized writers who claimed to see mythical or even religious scenes in stones, such as "Christ and Moses on the walls of the Baumann Cave; Apollo with the muses in the agate of Pyrrhus; the pope and Luther in the stone of Eisleben; and the sun, moon, and stars in marble." Leibniz considered these as "games not of nature, but of the human imagination."[39] Finally, images that looked just like real creatures, like "the coppery shapes of fish upon schistous stone . . . whose contours have been traced precisely, as if an artisan had inserted carved metallic material into the black stone," required a particular explanation.[40] They were the remains of actual animals. Therefore, they could be viewed as the documents of nature, that is, as evidence for a history of the earth.

Armed with this new "historical" perspective, Leibniz broke with the approach inspired by the hermetic and symbolist tradition of the Renaissance, of which Athanasius Kircher's *Mundus subterraneus* had given, as late as 1665, a particularly striking example.[41] In fact, Leibniz seems once to have been a follower of Kircher and to have accepted the "plastic virtue" of the earth and the "vegetation of stones." In an undated manuscript, originally written in French and recently discovered among Leibniz's papers in Hannover, he wrote:

> I find it hard to believe that the bones that now and then turn up in fields, or that are found while digging, are the remains of real giants; or that the stones of Malta, generally called snakes' tongues, are part

38. *Protogaea*, §II.
39. Ibid., §XXIX.
40. Ibid., §XVIII.
41. Kircher 1664, bk. 8, 53–62.

of fish. Or that shells found far from the sea are sure signs that these places were once covered by sea and that the shells were left behind when the waters receded, to petrify later. If this is so, the earth must be much older than the Bible indicates; but I propose by means of a rational process of reasoning, to show that this is not the answer. What I believe is that these shapes of animals and shells are usually nothing more than a game of nature: in other words, that they were created independently and have no relation to animals. For it is a fact that stones grow and take on many odd shapes; for proof of this we have only to look at the stones that R. P. Kircher accumulated in his Subterranean World.[42]

In this early text, Leibniz apparently denied the organic origin of "petri-fication" because of theological concerns. By the time he wrote *Protogaea*, though, Leibniz viewed certain fossil objects as organic remains; no mention was made here any more of the "plastic virtue of the earth."

In *Protogaea* Leibniz emphasized the astounding similarity between fossil objects and real animals. One could, for example, recognize different species of fish from their imprints. "I have here in my hands a barbel, a perch, a bleak, sculpted in stone. . . . I have also seen sea fish like the ray, the herring, and the lamprey, the last one sometimes lying crosswise with a herring." The impression of life that emerged from the sharp precision of these imprints was all the more convincing. "Not long ago an immense pike was dug out of a quarry, its body bent and its mouth open, as if it had been caught alive and turned to stone by the power of the Gorgon." To identify fossil objects as the remains of real creatures was to view them not as mere images, but as clues to a forgotten and buried history. Fossils provide us with an enduring language through which to reconstruct the past. This approach has much in common with Leibniz's interests in etymology, an interest reflected in *Protogaea* itself.[43]

Several related questions thus remained open: was it possible, as Bernard Palissy had surmised a century earlier, that certain species of animals had disappeared? Could the several varieties of "Ammon's horns," which Leibniz reproduced in the engraved plates accompanying *Protogaea*, have been "lost species"? Leibniz assumed instead that these species might exist in distant seas, or in unexplored ocean depths—perhaps at the bottoms of oceans—but he stressed that the world had been created

42. For a study of this manuscript, see Cohen 1998.
43. *Protogaea*, §XVIII.

complete, so that it was impossible to conceive of species that no longer existed. The "chain of being" had been created at once, as an immutable hierarchy. Though he sometimes entertained the notion that creatures had transformed themselves, gradually adapting to the environments in which they lived, Leibniz quickly pushed these notions aside as theologically unsound. "There are those who take the freedom to conjecture so far that they have imagined how once, when the ocean covered everything, animals that now live on land were aquatic; then, as the water departed, these animals became amphibians, until their descendants eventually left that original home."[44] Leibniz prudently added: "But that conflicts with the sacred writers, with whom it is impious to disagree." He must have been aware of the dangerously materialist character of this thesis, later systematized by "freethinkers," such as the French consul Benoît de Maillet, whose clandestine manuscript *Telliamed* asserted that all terrestrial animals had their ancestors in the sea, that birds came from fish, and that men and women came from mermaids.[45]

Leibniz did not calculate the timescale implied by his history, though he did recognize that if fossil shells were the true remains of animals, the earth itself must have been much older than the Bible claimed. In fact, there were prominent controversies about biblical chronology around 1700. Several freethinkers had dared to estimate the age of the world in hundreds of thousands, or even billions, of years.[46] Though Leibniz may well have been aware of these calculations, he did not take an explicit position about them in *Protogaea*.

Leibniz also shared with Steno the notion that human experiments could unmask natural processes. "I believe," wrote Leibniz, "that whoever more carefully compares the productions of nature with the fruits of the laboratories—that is indeed what we call the workshops of the chemists—will collect a reward, because an amazing similarity between natural and artificial things is often evident."[47] Steno wrote much the same thing in his 1667 essay on the dissection of a shark's head: "We owe

44. Ibid. §VI.

45. Benoît de Maillet's *Telliamed* was written between 1692 and 1720. It was distributed clandestinely after 1720 and printed for the first time in 1748 in Amsterdam. See Maillet 1748. On *Telliamed* and its possible relationship to *Protogaea*, see Cohen 2008.

46. See Marana 1710, bk. 3, 127. Some manuscripts of Benoît de Maillet's *Telliamed* place the earth's age at more than two billion years.

47. *Protogaea*, §IX.

these experiments to Chemistry, and I do not doubt that Nature operates in a similar way in the bosom of the Earth." As he argued for the organic origin of petrified teeth, Steno stressed the analogy between nature's underground laboratories and "art performed above the Earth." [48] This image of nature's workshop pervades *Protogaea* as well. Leibniz referred repeatedly to the ways in which human art recapitulated terrestrial history.

Skillful artisans could ape nature in the production of cinnabar or orpiment. Similarly, one could comprehend the petrification of living creatures through human art, which, like divine creation, used the properties of natural bodies in its own operations. If Leibniz rejected alchemical notions about the gestation of minerals, he nevertheless drew parallels between the hidden operations of nature and the technical processes of human art. For example, he compared the process of fossilization, in which heat and time combine to leave the imprints of fish on sedimentary rock, to the technique of reproducing insects out of silver. "We find something similar in the art of the goldsmith, for I gladly compare the secrets of nature with the visible works of men. They cover a spider or some other animal with suitable material, though leaving a small opening, and bake this material to stone in the fire. Then, by pouring in mercury, they drive the animal's ashes out through the hole, and, finally, they pour silver in the same way. When the shell is removed, they uncover a silver animal, with its entire complement of feet, hairs, and fibers, which are wonderfully imitated." [49] *Protogaea* thus stressed the human ability to reproduce nature's products by using nature's methods, thereby suggesting that one might reconstruct the logic of the past through experiment in the present.

The chemical workshops of Leibniz's time represented one especially concentrated site of experimentation dedicated to reproducing nature's products. And although he frequently criticized the superstition of simpletons, Leibniz certainly saw a useful purpose not only for productive chymists, but even for the Holy Roman Empire's many chemical projectors and charlatans: by aping nature, they provided greater insight into her methods. Leibniz himself had demonstrated early interest in alchemy while he was still a student at Altdorf, having served as the secretary of

48. Steno [1667] 1969, 112–113.
49. *Protogaea*, §XVIII.

a secret alchemical society in nearby Nuremberg.[50] He did not stay there long, and he treated the society's activities with some skepticism, but it would be unfair to claim that Leibniz rejected alchemy, for he traveled in the same circles as alchemists like Johann Joachim Becher and Johann Daniel Crafft, and he retained a serious interest in the subject.[51] Nevertheless, though Leibniz stayed abreast of developments in chemistry—as in almost every other field of knowledge—it does not seem that he did much laboratory work, and *Protogaea* does not contain evidence of original chemical experiments or discoveries. What the work does demonstrate, however, is his strong interest in the chemical literature of his time and his willingness to gather new information from travelers, smelting masters, and even swindlers. The chemical laboratory, he believed, could experimentally re-create the world in miniature, and therein lay its attraction. As he put it in *Protogaea:* "nature, using volcanoes as furnaces and mountains as alembics, has accomplished in mighty works what we play at with our little models."[52]

From Natural History to Human History

Leibniz arrived in Hannover in 1676 as official court librarian and counselor to Johann Friedrich, the duke of Brunswick. Though he encountered some unpleasant court politics during his first years in Hannover, Leibniz eventually secured himself a good situation there. In 1680, he committed to write a history of the House of Brunswick, but it would never appear during his lifetime. As historian for the dukes of Hannover, Leibniz had to establish the legitimacy of their dynastic claims, and, as part of this charge, he traveled extensively between 1687 and 1690 in Germany and Italy to look for sources in archives and libraries.[53] Leibniz's efforts did result in the publication, between 1701 and 1711, of treatises and docu-

50. See Ross 1984, 5–6; 1974.

51. See Smith 1994a; and Ross 1974.

52. *Protogaea*, §X. It would be misleading to regard Leibniz either as a precursor to modern chemistry or as a disciple of the alchemists. Writing at the end of the seventeenth century, he did not clearly distinguish between chemistry and alchemy. Rather, it would be better to speak about "chymistry," a rich field of knowledge and practice that relied on techniques from the alchemical laboratory while producing everything from pharmaceuticals and dyes to saltpeter and perfume. See Newman and Principe 2004, ix–xvi.

53. See Robinet 1988.

ments related to the House of Brunswick. His method consisted of gathering testimonies and official documents, classifying them, verifying them, and discussing their authenticity in order to trace origins.[54]

Leibniz intended *Protogaea* as a preface to his monumental history of the House of Brunswick, the *Origines Guelficae*, but this work remained unfinished and unpublished during his lifetime. Thanks to the efforts of Christian Ludwig Scheidt, it finally appeared in print around 1750, shortly after *Protogaea*. The links between human and natural history were more than incidental, for Leibniz employed the methods of a historian in *Protogaea*, gathering "documents" from strata and caves as if from the "archives of nature." He collected glossopetrae, "Ammon's horns," and bones from cave deposits. *Protogaea* enumerates and describes the objects discovered underground and in the caves of Lower Saxony. Its plates represent and reconstruct these objects. The task of the historian, as Leibniz saw it, involved gathering these scattered documents, remnants of the earth's ancient past. The present appearance of a region thus provided clues about a time before memory and chronicle. *Protogaea*, then, dealt with "the first face of the earth" and studied the remains of its most ancient history.

The challenges of reconstructing a general history of the earth were, in some ways, similar to the challenges of writing human history. During this same period, historians debated about whether to emphasize the establishment of historical evidence or the construction of narrative. In France, for example, the *Querelle des anciens et des modernes* debated the character and construction of history. "Antiquarians" distanced themselves from literary forms of history, publishing inventories, sources, and chronologies that did not aim at universal narratives. The great Maurist compilations were built on this foundation.[55] Others focused on antiquity using the same methods. Bernard de Montfaucon, a paleographer and philologist, provided illustrations of a wide range of objects from antiquity in his *Antiquité expliquée et commentée en figures*.[56] These included representations of gods, religious practices, and material life.[57]

54. On the political context of Leibniz's historical work, see Meyer 1952; and Daville [1909] 1986.

55. These included the *Acta sanctorum ordinis sancti benedicti* of 1688, the *Annales ordinis sancti benedicti* of 1703, and the *Gallia christiania* of 1715. Cf. Barret-Kriegel 1988.

56. Montfaucon 1719.

57. Schnapp 1993, 236.

In contrast to such erudition without narrative, narrative without erudition had as its aim the celebration of the sovereign. This "monumental" history, which used the epic style, was written by men of letters (Jean Racine and Nicolas Boileau were appointed historiographers to the king in 1677) and was dominated by moral concerns and the demand for eloquence. François Fénelon, for example, illustrated the relationship between history and literature at this time, distinguishing the historian from the compiler, that is, from the author of a history "chopped into little pieces without any benefit from a lively narrative, which could join them together" and from the "dry and sad record maker," who "cannot see anything but chronology."[58] History, according to Fénelon and the moderns, should be primarily concerned with constructing a narrative, which in turn linked it to fiction.

Bernard de Fontenelle, for his part, emphasized the difficulties created by the construction of a historical *system* of the formation of the earth, which could end up being even more dubious and unreliable than a philosophical system. Critical of the chronologists' attempts, he emphasized the difficulties posed by any effort to reconstruct the past from the inevitably partial traces of the present:

> Let us assume that we were able to find the fragments from the ruins of a huge palace, which have been scattered over a large area of land. Should such be the case, and if we were certain no piece was missing, it would be a prodigious work to collect them all, or to come up with an accurate idea of the entire structure of this palace just by looking at them. Moreover, should there be some pieces missing, it would be even harder to imagine the structure. Indeed, the more fragments that were missing the more difficult it would be. Under these circumstances we would be likely to produce different plans of the building that had almost nothing in common with the original. This is the state, at present, of our most ancient history.[59]

Indeed, the waning years of the seventeenth century witnessed a tension between fragmentary and universal accounts in the narratives about the formation of the earth. Though some diluvialists succeeded in giving a complete narrative of terrestrial history, others presented partial histo-

58. Fénelon [1710s] 1897, 79–90.
59. Fontenelle [1735] 1766.

ries. *Protogaea* is itself a fragmentary text, more like a succession of notes than a coherent systematic exposition. It dwells on the identification of fossils, the description of springs and curious mineral formations, and the legends of the *Nibelungen*. It reflects the tension between the criticism of local "documents" (fossils, curiosities) and the systematic attempt to construct a global narrative.

Leibniz's history of the earth was, by his own admission, partial and fragmentary. But he imagined a more complete history, in which correspondents from around the globe would pool local knowledge in an effort to create a more accurate whole. In fact, Leibniz intended *Protogaea* as a model for others, since "when everyone contributes curiosity locally, it will be easier to recognize universal origins."[60] The relationship between part and whole, particular and universal, fragment and totality played a central role in Leibniz's metaphysics.[61] In the same way, one small region of the globe, like Lower Saxony, might mirror the general history of the earth.

A New Science Called "Natural Geography"

In the preamble to *Protogaea*, Leibniz writes that "our homeland is the source of remarkable speculations, and the rays of a public light emanating from here will also advance the exploration of other regions." Much of the work's originality stems from this opening commitment, as Leibniz labors constantly to situate an abstract theory of the earth within the material circumstances of one particular region. Steno's *Prodromus*, it is true, was also like this, but Steno limited himself to a succession of earth layers in Tuscany. Leibniz, on the other hand, went much farther, by exploring the Harz region in all its particularity. He not only wrote about its appearance, terrain, mineral wealth, and mining practices, but also showed interest in its curiosities, local legends, histories, and even the etymology of local names.

Local knowledge lies at the heart of Leibniz's "new science called natural geography," which he announces near the beginning of the work.[62] Though rooted in sweeping causal deductions about the development of the "great bones of the earth," Leibniz's natural geography

---

60. *Protogaea*, §I.
61. Cf. Leibniz 1989, 221 (*Monadology*, §62).
62. *Protogaea*, §V.

rested on myriad particular observations, like the structure of ore veins at Osterode or Ramelsberg, the petrifications of Eisleben and Lüneburg, or even the superstitions of miners in the Harz. Leibniz had observed many of the curiosities and particulars described in *Protogaea* himself. In his account of the Baumann and Scharzfeld caves, for example, we go spelunking with him and share in the wonders of the adventure. For Leibniz, natural geography had to rest on the solid foundations of painstakingly accumulated local knowledge.

Consider, for example, Leibniz's reflections about fish imprinted on slate.[63] The main point here, it would seem, involved demonstrating the organic nature of fossil fishes. But Leibniz does much more than that, referring to the "ichthyomorphic stones" found in Eisleben, "a Saxon town in the region of Mansfeld, near our Harz town of Osterode"; he lingers over the "black foliated stone . . . properly called slate"; he pauses to consider the Latin word *Ardosia*, its German equivalent, *Lagen*, and its Upper German variant, *Laya*. We then discover that *Laya* "is also the common name of the famous de Petra family, which (a very rare example) in our time simultaneously produced two brothers as electors along the Rhine." *Protogaea* is filled with strange, apparently rambling passages like this one.

This attention to detail, and the gift for finding universal relevance in the smallest local features, prompted Leibniz to dwell on particular specimens and specific sites, especially when he had personal experience of them. He thus grounded the analysis of ichthyomorphic stones with a personal reflection:

> I own a fragment of such a stone, each side of which is imprinted with the image of a different fish. They are found in a hanging vein, since after excavation of the superficial clay earth and the subsequent rocks, there occur in Eisleben various layers of coppery schist. But only one layer has fish, and this one is especially well suited to the fire, for no other copper ore obeys the smelter more easily. This layer is about sixteen inches thick, though sometimes it gets as thin as a knife blade; but the narrower the mass, the richer it gets. The vein is enclosed on each side by walls of the hardest stone.

Leibniz's method allowed him to move beyond the isolated specimen, displayed disconnectedly in some cabinet of curiosities. Here, by con-

63. Ibid., §XVIII.

trast, is a full, concrete, detailed account of the petrified fish together with an account of its *situation*. Natural geography demanded thousands of local observers, each contributing detailed, situated knowledge about particular things.

The productive tension between particular and universal also manifested itself in the striking blend of literature that Leibniz drew upon in constructing his argument. Leibniz often moved abruptly from the most abstract considerations to the smallest local details—Cartesian metaphysics to Harz homunculi. Some of the more prominent European figures, like Steno, Descartes, and Burnet, left their mark upon *Protogaea*. But the local works of more obscure German naturalists, like Valerius Cordus and Friedrich Lachmund, were no less important.[64] Still, it was Agricola who combined an international reputation with detailed local knowledge.

In fact, though Leibniz and Agricola were separated by more than a century, they shared some striking similarities. Both men hailed from the German silver states and traveled often to Italy, and both had personal experience of the great German mining districts that inspired their writings. It should come as no surprise, then, that Leibniz relied on Agricola's writings as he turned his attention to mines and minerals. If Agricola's posthumous *De re metallica* (1556)—provided an essential source on mines and mining, his earlier work *De natura fossilium* (1546) was more important for *Protogaea*.[65] For it was in this latter book that Agricola turned his attention to minerals, "congealed juices," metals, "earths," stones, and in general anything "that people dig up."[66] It was also in *De natura fossilium* that he established a technical vocabulary for mines, metals, minerals, and fossils that would persist for several centuries. Though Leibniz took issue with many of the classifications, descriptions, and explanations inherited from this work, *Protogaea* shows the lasting impact of its terminological structure.[67]

64. See Ogilvie 2006, 145–150.

65. Froben published several of Agricola's works in Basel, together with the second edition of *Bermannus*, in a single folio volume in 1546. The first of these was *De ortu & causis subterraneorum*, in five books. *De natura fossilium*, in ten books, appears later in the same folio volume. The first English translation of *De natura fossilium*, by Mark Chance Bandy and Jean A. Bandy, appeared in 1955.

66. Agricola 1546, 1.

67. Comparison with *De natura fossilium* (1546) reveals the terminological continuity between Agricola and Leibniz.

But if Leibniz owed much to *De natura fossilium*, he did not simply accept Agricola as an authority. Leibniz seems to have treated Agricola's work like any other source, that is, as something subject to criticism, comparison, and investigation. In several cases, after consulting other authors or visiting the sites himself, Leibniz concluded that Agricola was mistaken. For example, Leibniz argued that there never had been alum mines of consequence in Lüneburg, as Agricola and his followers liked to claim. He even called into question Agricola's physical description of the area: "Near Lüneburg, at the foot of a mountain upon which a brickworks has been built, the earth is salty or aluminous and not as rich as Georgius Agricola described it. Instead, it is thin, almost sandy, and thus unsuitable for making bricks, unless one digs the earth out of deeper shafts, exposes it to the rain and sun, and moistens it properly until it toughens."[68] Leibniz also relied on his own observations and on Steno's work to criticize Agricola on a number of issues, like the origin of glossopetrae, the smell of ostracites, and the character of petrified wood.[69]

Leibniz believed that "Agricola received most of his knowledge about our fossils in the quarries, wells, graves, and cellars of Hannover, and especially of Hildesheim," from Valerius Cordus, "the distinguished physician of Brunswick and Hildesheim."[70] He also relied on another Hildesheimer, Friedrich Lachmund, whose 1669 *Oryktographia Hildesheimensis* provided detailed accounts and descriptions of local discoveries, like trochites, miraculous marble images, osteocolla, and Ammon's horns.[71] It contained a series of plates illustrating the fossils, stones, and other terrestrial wonders of Hildesheim, which Leibniz wanted to include in *Protogaea*. All editions (starting with Scheidt's in 1749) have included both a series of plates and extensive pieces of text from the *Oryktographia*.[72] Interestingly, *Oryktographia* is filled with the kind of miraculous and superstitious claims that *Protogaea* sought to dismiss, like seeing religious images in rocks and caves. Here, as with Agricola's work, Leibniz showed himself the pragmatist, using what he could and discarding the rest. In fact, when it came to the Harz and its surrounding region, Leibniz valued the

68. *Protogaea*, §XXX.
69. Cf. ibid., §XXX, §XXXI, and §XLV.
70. Ibid., §XXIII.
71. Friedrich Lachmund (1635–1676). The full title was *Oryktographia Hildesheimensis, sive, Admirandorum fossilium.*
72. See below, xxxix–xl, 96–97, and the appendix.

authority of *local* witnesses. Given *Protogaea's* emphasis on the importance of local knowledge, this should come as no surprise.

Manuscripts and Editions

Leibniz wrote *Protogaea* between 1691 and 1693. But it did not appear in print until 1749, when Christian Ludwig Scheidt rescued it from the "moths and the dust" of the Royal Library in Hannover, where it had been "consigned to a long oblivion" under piles of paper and books. Scheidt, following in Leibniz's footsteps, had been appointed historian and librarian for the House of Hannover, and it was in this capacity that he obtained exclusive permission to "edit all of Leibniz's unpublished works for publication." Interestingly, with a staggering wealth of material at his disposal—most of Leibniz's writings were unpublished in 1749—Scheidt turned to *Protogaea* first. "This treatise is well suited to the efforts of a historian," he explained, "and I consider it a worthy text with which to inaugurate the office that has been conferred upon me."[73]

Ironically, Scheidt did not much like *Protogaea*. He rejected the notion that violent upheavals and catastrophes had created the mountains and the seas; he denied that the Great Flood could be attributed to natural causes; and, above all, he disavowed Leibniz's "conjectural" approach to history. Scheidt considered Leibniz's leaps of reason, through which he tried to trace terrestrial history back to its origins, impious. It would be better, Scheidt counseled, to follow the modest path of Augustine and confess our ignorance, rather than allow cleverness to lead us astray. In short, when *Protogaea* appeared some five decades after Leibniz had written it, it came with a disclaimer: though interesting, the book's method was flawed and dangerous.

In fact, *Protogaea* was, quite literally, only the beginning. "For if things had gone according to Herr Leibniz's will," Scheidt explained, "then this treatise would have been the preliminary effort in a much larger work." Leibniz's immediate successors, Johann Georg Eckhart and Johann Daniel Gruber, had eventually brought the monumental work to conclusion. Now Scheidt intended to produce a luxurious edition worthy of its subject. When *Protogaea* appeared in 1749, it was as a preface to the massive history of the House of Brunswick that Leibniz had initiated some seven

73. Leibniz 1749, v–vi, xiii.

decades earlier: the *Origines Guelficae*.[74] The physical appearance of the various editions reveals Scheidt's priorities. His 1749 edition of *Protogaea* was a slim, relatively unassuming volume (even with its appended copper engravings); the three volumes of the *Origines Guelficae*, on the other hand, appeared in three massive, costly, sumptuous volumes between 1750 and 1753.

Though *Protogaea* never appeared in print during Leibniz's lifetime, he did submit for publication three or four shorter papers on similar subjects between 1693 and 1710. The most famous of these was the first, "Protogaea," which appeared in the *Acta eruditorum* in 1693.[75] The second, a letter of 1697 to the Royal Society in London, may have been intended for the *Philosophical Transactions*.[76] A transcription of the third appeared in abridged form in the Paris Academy's *Histoire* of 1706.[77] The last, a commentary on Philipp Jakob Spener's fossil "crocodile," was published by the Berlin Academy in 1710.[78] Moreover, Leibniz included several paragraphs on the history of the earth in his *Theodicy*, published in the same year.[79]

Thus, after 1693, Leibniz distributed many extracts of his text, and his views also circulated through his considerable network of correspondents, which included Louis Bourguet and Bernard de Fontenelle. This circulation of summaries and excerpts meant that *Protogaea*'s theses became well known not only among German scholars, but also in France, long before its actual publication. It should not be surprising, then, that Leibniz's ideas would become a major source for Georges Buffon's 1749 work *Theorie de la terre*.

Scheidt's edition of *Protogaea*, which appeared in that same year, included a series of illustrations, and we know that Leibniz had organized and arranged these for publication with the text. Nicolaus Seelander, official engraver for the Royal Library in Hannover, who had been at Leibniz's disposal to illustrate the *Origines Guelficae*, made these copper engravings. Seelander's illustrations were not abstract sketches of the earth and the phases of its transformation—the kind of thing we see in Descartes's *Prin-*

74. Scheidt 1750–1753.
75. Leibniz 1693.
76. Cf. Sticker 1967.
77. Rappaport 1997a, 6–11.
78. Leibniz [1710] 1966. See also Rappaport 1991 and 1997b.
79. Leibniz 1710, §242–§245.

cipia. Rather, the first engraving in *Protogaea* reveals a cross section of the Baumann Cave, and it indicates where individual fossils and mineral formations were found (see fig. 14). As in other contemporary works dealing with fossil objects, *Protogaea*'s copper plates represented "petrifications," and these were meant to offer various proofs or arguments about terrestrial history. The plates presented, among other things, Ammon's horns, fish imprinted on slate, and different shells arranged according to their shapes. While some of the engravings reproduced objects from the library's collections in Hannover, others attest to the wide circulation of images during this period. Figure 7, for example, reproduces an engraving from Steno's essay on the dissection of a shark's head.[80] Steno took his illustration, which sought to demonstrate the organic origin of glossopetrae by representing them in the shark's mouth, from Michele Mercati's unpublished *Metallotheca vaticana*.

But the single most important source for illustrations in *Protogaea* was Friedrich Lachmund's *Oryktographia Hildesheimensis* (1669), which provided material for six of the twelve plates that Leibniz planned to include in his work (figs. 4, 5, 8, 9, 10, 11). The illustrations in *Oryktographia*, based on specimens from Lachmund's own collection, were really nothing more than sketches.[81] Interestingly, the engravings for *Protogaea* were of much higher quality than the originals. Did Leibniz, or his engraver, have direct access to Lachmund's collection? We have found no evidence for that. Moreover, since the engravings for *Protogaea* retain the same labels as the sketches in *Oryktographia*, it seems likely that they were copied directly from Lachmund's book.

Figure 12 in *Protogaea* is a reconstructed "unicorn" skeleton, together with the "tooth of a marine animal." This illustration has been sometimes viewed as the first vertebrate reconstruction in the history of paleontology (albeit a curious one): it seems to be composed of fossil proboscid and rhinoceros bones, and the tooth is that of a mammoth.[82] In describing the remains of this animal, Leibniz explained that his description was borrowed from Otto von Guericke, mayor of Magdeburg and inventor of the air pump. In the same treatise that described experiments with the air pump, Guericke recounted the discovery of the unicorn skeleton.

80. Steno [1667] 1969.
81. See Lachmund 1669, 41; and the appendix.
82. See Abel 1925; and Cohen 2002. This engraving appeared as plate 12 in Scheidt's 1749 edition.

The remains of the unicorn had been discovered in a gypsum quarry near Quedlinburg in 1663, "with the rear part of its body bent back, as is common with animals, but with a raised head and carrying on its forehead an extended horn about five yards long." [83] But this image does not appear in Guericke's book, and it is likely that Leibniz had his engraver reproduce and improve upon an existing drawing, a sketch that had circulated in contemporary periodicals. Though the unicorn may seem fanciful to us, Leibniz considered it a sound visual argument for his history of the earth, an argument authenticated by the authority of its witness.

Eduard Bodemann's 1895 catalog of Leibniz's papers mentioned two manuscript versions of *Protogaea*. The "A" version, which Leibniz revised himself, served as the basis for Wolf von Engelhardt's 1949 German translation. The "B" version, later revised by Eckhart, served as the source for Scheidt's 1749 Göttingen edition, Bertrand de Saint-Germain's 1859 French edition, and Jean-Marie Barrande's 1993 French edition. Though the A manuscript (the basis for our translation) has survived, the original B manuscript was lost during World War II, and all that remains of it is Eckhart's revision. [84]

The structure of the 1749 Latin edition of *Protogaea*—and all subsequent editions of the work—owed much to the editorial work of Leibniz's successors, Eckhart and Scheidt. Eckhart's revised B manuscript (Ms XXIII, 23b), for example, demonstrates the considerable impact of their revisions. It was probably Eckhart who inserted new text and images (especially from Lachmund) into the existing manuscript; after that, Scheidt divided the work into numbered paragraphs with subtitles. The result, with its discrete sections and chapter headings, somewhat masked the fragmentary and disconnected flavor of the original manuscript. [85]

We have adopted Scheidt's division and numbering of sections, which reflects the paragraph structure of the original. We have also followed Scheidt in providing headings for each section, but these headings, which here appear only in English, have been bracketed to indicate their absence in the original. Moreover, some of the sketches that appear in Leibniz's A manuscript have been reproduced here. We have also interleaved the

83. *Protogaea*, §XXXV; Guericke 1672.

84. Gottfried Wilhelm Leibniz Bibliothek, Hannover, Handschriftenbestand, Ms XXIII, 23a and 23b. Ms XXIII, 23a, includes the so-called "A manuscript." Ms XXIII, 23b, features Eckhart's revised version of the original B manuscript.

85. Leibniz 1749, xxxiv–xxxvi.

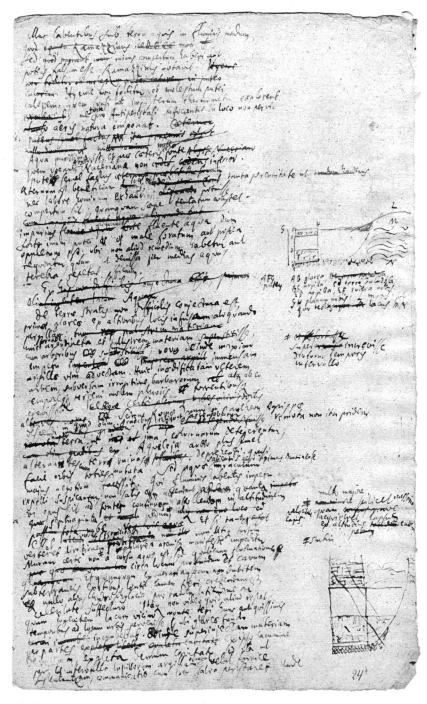

FIGURE 2

Page from the "A manuscript" of *Protogaea*, in Leibniz's hand, with original sketches. See *Protogaea*, §VIII, on the wells of Modena. (Gottfried Wilhelm Leibniz Bibliothek, Hannover.)

engravings with the text and placed them adjacent to relevant paragraphs, rather than simply appending them to the end of the work, as in other editions. This reflects our conviction that these illustrations are central to the argument of *Protogaea*. Moreover, the revised B manuscript had very clear indications about where to insert the various plates, but Scheidt's 1749 edition ignored these instructions and put the illustrations at the end of the work. Finally, we have taken the text from Lachmund's *Orykto-graphia*, which Eckhart inserted directly into the body of *Protogaea*, and moved it to the appendix.

Claudine Cohen and Andre Wakefield

PROTOGAEA

[I]

Magnarum rerum etiam tenuis notitia in pretio habetur. Itaque ab antquissimo[1] nostri tractus statu orsuro dicendum est aliquid de prima facie terrarum, et soli natura contentisque. Nam editissimmum[2] Germaniae inferioris locum tenemus, maximeque metallis foecundum; et domi nobis insignes conjecturae, et velut radii nascuntur publicae lucis, unde ad caeteras regiones aestimatio procedat. Quodsi minus assequimur destinata, saltem exemplo proficiemus: Nam ubi in suo quisque curiositatem conferet, facilius origines communes noscentur.

[II]

Globum terrae, ut omnia nascentia, regulari forma e naturae manibus exiisse sapientibus placet: Nam nec deus incondita molitur;[3] et quicquid per se formatur, insensibiliter aut concrescit per particulas, aut pro sese disponentium delectu conflictuque tornatur. Itaque asperitas montium, quibus horret facies orbis, postea supervenit. Et certe si liquidus initio fuit, etiam aequabilis fuerit;[4] generalibus autem corporum legibus consentit, firma ex liquidis induruisse. Quod et solida intra solidum clausa testantur, stratis quibusdam nucleisque in suos angulos limitesque persaepe decircinatis, venae in rupibus, gemmae in saxis. Sed et rerum veterum spolia passim extant, plantarum, et animalium, et arte factorum, sub novo et lapideo involucro. Itaque ambiens quod nunc durum cernimus, postea natum est; tunc vero adhuc fluidum fuisse oportet. Porro ipsa fluiditas ab intestino est motu, et tanquam gradu caloris; quod indicant experimenta: Nam imminuto calore etiam aqua in glaciem consistit; dum contra corrodentes liquores, et ab occulto motu fortes, difficulter

---

1. B: antiquissimo. ("B" denotes variations that appear in the B manuscript but not in the A manuscript.)
2. B: editissimum.
3. B: Deus enim incondita non molitur.
4. B: aequabilis fuerit necesse est.

[1. *Preamble*]

Even a slight notion of great things is of value. Therefore, those who would trace our region back to its beginnings must also say something about the original appearance of the earth, and about the nature of the soil and what it contains. For we occupy the highest region of lower Germany, one that is especially rich in metals.[1] Moreover, our homeland is the source of remarkable speculations, and the rays of a public light emanating from here will also advance the exploration of other regions. But if we do not completely achieve our goal, then we will at least have a model, for when everyone contributes curiosity locally, it will be easier to recognize universal origins.

[11. *The first formation of the earth through fire*]

The philosophers like to argue that the globe of the earth, like everything that has been born, arose in regular form out of the hands of nature. For God makes nothing without order, and everything that forms itself develops imperceptibly out of small parts, or is shaped by the separation and collision of these parts. That is why the jagged mountains, which bristle on the face of the earth, appeared only later. Certainly, if the earth was liquid at the beginning, it would have been uniform; and it agrees with the general laws of bodies that solids harden out of liquids, as witnessed by solids enclosed in solids, and by layers and kernels enclosed in the fissures of the earth, like veins in rocks and gems in stones.[2] And then there are the scattered vestiges of old things, of plants, animals, and artifacts wrapped in a new coat of stone. Thus, what we now perceive as hard appeared only later and must certainly have been liquid once. Now this same fluidity arises from an inner movement and a certain degree of heat, as is indicated by experiments. For in the presence of reduced heat, water hardens to ice, while, in contrast, acidic liquids and those animated by a hidden motion harden with difficulty. Furthermore, heat

---

1. The Harz Mountains, where Leibniz served as a mining consultant for the dukes of Brunswick. See introduction, "Leibniz in the Harz."

2. Cf. Steno 1669. See "Introduction," xxv–xxvi.

congelantur. Calor autem motusve intestinus ab igne est, seu luce, id est tenuissimo spiritu permeante. Atque ita ad motricem causam perventum est, unde Sacra quoque Historia Cosmogeniae initium capit.

[III]

Quousque ergo pertingere hominum notitia potest sive ratiocinatione, sive[5] scripturarum traditione, primus est formationis rerum gradus, separatio lucis et tenebrarum, id est agentium et patientium; secundus patientium inter se discriminatio, id est liquidorum discessio a siccis, quae duo distinguuntur pro diversa in patientibus resistendi facultate et gradu firmitatis. Itaque incendiis et inundationibus varie transformata sunt corpora. Et quae nunc opaca et sicca cernimus, arsisse initio, mox aquis hausta fuisse, tandemque secretis elementis in praesentem vultum emersisse credi par est. Quibus consentanea quidam sapientiae mystae statuunt in hypotheseos formam, distinctiusque explicant separandi modum. Nempe globos quosdam mundi ingentes cum ad fixae stellae, aut nostri solis modum per se lucerent, aut ex sole suo ejecti essent, mox excocta ac spumescente materia, exurgentibus a fusione scoriis, fuisse obductos; veluti si sol maculis invalescentibus velaretur, quibus infici aliquando eum, quin et obscurari subagnoscebent veteres, nostra aetas reperta oculi armatura pervidit. Excessu autem collectae materiae fractus calor internus, et crusta in ambitu refrigerata consistebat. Inde nascebatur opacum sidus jam alienos radios remissurum, ut planetae. Talem Vulcanum nos habitare vel suspicantur vel fingunt, Mosaico illo lucis ac tenebrarum divortio factum. Sane plerisque creditum, et a Sacris Scriptoribus[6] insinuatum est, conditos in abdito telluris ignis thesauros, aliquando iterum erupturos. Adjuvant conjecturam extantia adhuc vestigia primi naturae vultus. Nam omnis ex fusione scoria vitri est genus; scoriae autem assimilari debuit crusta, quae fusi globi materiam[7] velut in metalli furno obtexit, induruitque post fusionem. Talem vero esse globi nostri superficiem (neque enim ultra penetrare nobis datum), reapse experimur. Omnes enim terrae et lapides igne vitrum reddunt, sed tanto

---

5. B: sive Sacrarum Scripturarum propagatione ac traditione.
6. B: sacris etiam Scriptoribus.
7. B: fusam globi materiam.

and inner motion come from fire or from light, that is, from a very subtle and penetrating spirit. And so we have arrived at the motive cause which sacred history takes as the beginning of cosmogony.

[III. *Different opinions concerning the creation of the globe*]

Insofar then as it is possible for human knowledge to reach back, whether through reasoning or through the tradition of the scriptures, the first step in the formation of things is the separation of light from darkness, that is, of the active from the passive. The second step involves the differentiation of passive things from one another, that is, of the wet from the dry. Wet and dry things, in turn, are separated from one another by their power of resistance and degree of firmness. Bodies are therefore transformed by fires and waters in different ways. In all likelihood, those that now seem opaque and dry were initially ablaze; then they were swallowed by the waters; and finally, after the separation of elements, they assumed their present appearance. This conforms to what certain priests of wisdom have constructed, in the form of hypotheses, to explain more distinctly how such a separation of elements might have occurred.[3] Indeed, they suggest that there were once huge globes, like the fixed stars or our own sun, that either produced light or were jettisoned by a sun. Then their matter boiled and foamed until they were finally covered by the slags extruded during fusion.[4] Similarly, as the ancients supposed, the sun would be veiled by increasing numbers of spots that would darken and eventually obscure it, something actually observed in our time, after the invention of the armed eye.[5] Still, the accretion of accumulated material extinguished the internal heat, with a cooled crust hardening all around. Thus was born an opaque star that would reflect external rays, just like the planets. They either suppose or imagine that we inhabit a volcano fashioned, as Moses wrote, through the division of light from darkness. Certainly, most believe, as is suggested by the sacred scriptures, that there are secret chambers of fire buried deep in the earth that will

---

3. Cf. Descartes [1644] 1983, 181–182 (IV, §2–3); Leibniz 1960, vol. 3, 565–566.

4. Leibniz here draws on well-known processes of smelting for base and precious metals. Cf. Agricola [1556] 1912, 222–223.

5. In other words, after the invention of the telescope.

magis, quanto propius ad rudis saxi speciem accedunt. Neque interim negaverim, altius productis transformationibus posse terrestria et vitrescentia gigni ex aquis, quas variis corporibus foetas esse constat; ipsaque materies per se ubique similis sibi quamcunque formam induere potest; neque ulla sunt ultima incommutabiliaque elementa. Sed nobis hoc loco satis est, admoto, humana arte, efficacissimo agentium igne, terrestria in vitro finiri. Ipsa magna telluris ossa, nudaeque illae rupes, atque immortales silices, cum tota fere in vitrum abeant, quid nisi concreta sunt, ex fusis olim corporibus a prima illa magnaque vi, quam in facilem adhuc materiam exercuit ignis naturae? Is enim nostrorum furnorum efficaciam immenso gradus durationisque excessu superans, quid mirum est, si tunc produxit, quae nunc homines imitari non possunt: quanquam et ars quotidie proficiat, et subinde res novas inauditasque proferat, immo etiam ad magnam aliquando duritiem provehat corpora igne suo fusa. Cum igitur omnia, quae non avolant in auras, tandem fundantur,[8] et speculorum inprimis urentium ope vitri naturam sumant, hinc facile intelligas, vitrum esse velut terrae basin, et naturam ejus sub caeterorum plerumque corporum larvis latere, particulis varie corrosis atque subactis, partim solutione agitationeque aquarum, partim repetitis in vapore elevationibus et destillationibus, donec accedente salium opera ad vim caloris, in limum corrumperetur saxea durities, alendis plantis et animalibus convenientem, et in volatilem quoque naturam eveheretur. Interim quo quidque in tellure magis nudum aut primitivum est, et compagi ipsi rupium affine, hoc magis persistit in igne, summoque calore funditur, et postremo vitrescit. Nam et calcarius lapis, qui furnis resistit, speculo in vitrum domatur. Quin et arena, quae magna, et simplicissima pars terrae est, immensaque deserta, et littora, et fundum maris opplet, et meliori solo glaream substernit, propius intuenti quid aliud refert, quam lapillos, seu fluores perlucidos, et velut vitrum, motu aut in ipsa fusione, aut alias comminutum? Quod et facili ignis opera reddit, si sales accedant, qui nec initio defuere.

8. B: funduntur.

erupt again sometime.[6] They support this conjecture through the existing vestiges of nature's first face. For all slags produced through fusion are a kind of glass. But the crust itself must have been like a slag that covered the molten earth mass, as if in a blast furnace, eventually hardening after fusion. We actually experience that the surface of our globe has been made in this way (and of course it is not granted us to penetrate any further). For all earths and rocks return to glass through fire, the more so as they approach the appearance of raw stones. Nor would I deny meanwhile that, through further transformations, earth and glass could be produced out of water, which is teeming with divers bodies, as everyone knows. This same material, which is everywhere identical with itself, can take on any form, since there are no ultimate, unchangeable elements.[7] But for us it is enough to note that, through human art and its most effective agent, fire, earth turns to glass. The great bones of the earth, naked rocks and immortal sands, almost all change to glass. How were they formed if not through the prior melting of bodies by that first great force, which worked the then still tender material with the fire of nature? This fire greatly surpassed the power of our own furnaces, both through a much longer duration and a much greater degree of heat. It is no wonder, then, if the fire of nature made something that people cannot imitate now, even though art progresses every day, continually producing new and surprising things, and sometimes even rendering the bodies melted in its fire extremely hard. Since, therefore, all things that do not fly away in the winds are eventually melted, and, particularly, since all things take on the nature of glass through the use of burning mirrors, it is easy to see that glass forms the basis of the earth and that its nature lies hidden behind the mask of most other bodies.[8] For it was divided and kneaded into little particles in a variety of ways, partly through the dissolving influence of moving waters, partly through repeated evaporation and distillation, until the action of salts was added to the force of the heat, thereby eroding the hard rocks to a fertile soil, which could nourish plants and animals; and this volatile element would also have been carried upward. At the same time, the more anything in the earth is naked or primitive and

6. Burnet 1681, bk. 3, chap. 3.

7. Cf. Leibniz 1960, vol. 3, 398–399; 1989, 132–133.

8. Burning mirrors were a subject of intense interest throughout the seventeenth century. Leibniz would have been acquainted with the famous "burning machine" of his friend Ehrenfried F. Tschirnhaus. See Tschirnhaus 1697.

Ex hac genesi rerum jam inobservata[9] hactenus procedit salis marini origo. Nam ut perusta, ubi refriguere, humorem attrahunt, unde olea per deliqium[10] Chemicis nascuntur in cella; ita pronum erit credere, sub rerum initiis, nondum separato a luce opaco, cum globus noster adhuc arderet, pulsum ab igne humorem abiisse in auras, deinde verum destilationum exemplo renatum, mox remittente aestu in aquosos vapores iterum fuisse densatum, et cum a congelascente terrestris superficiei massa resorberetur, in aquam denique rediisse, quae terrae faciem abluens vasta recentis empyreumatis vestigia salemque fixum in se recepit. Unde natum est lixivii genus, quod deinde in mare confluxit. Sane ex plantarum analysi[11] compertum habemus, duo salis fixi genera in lixiviis restare, alterum alcalicum, ut loquuntur artifices, ducta voce ab herba, quam nostri sodam, Arabes *Cali* appellant, alterum marinum, magisque ad acidum inclinantem. Postremo credibile est, contrahentem se refrigoratione crustam, ut in metallis, et aliis, quae fusione porosiora fiunt, bullas reliquisse, ingentes pro rei magnitudine, id est, sub vastis fornicibus cavitates, quibus inclusus fuit aer humorve; tum etiam in folia quaedam discessisse, et varietate materiae calorisque inaequaliter subsedisse massas, quin et dissiluisse passim fragminibus in declivia vallium inclinatis, cum partes firmiores, et velut columnae, supremum locum tuerentur;

9. B: observata.
10. B: deliquium.
11. B: ut jam in Parisiensium Academicorum observationibus notatum est.

related to rocks, the more it resists fire; only the highest degree of heat can melt it, and ultimately turn it to glass. For even limestone, which resists the ovens, changes to glass by means of burning mirrors. Yes, even the sand, an important and indeed the simplest part of the earth, which fills the deserts and the beaches and the bottom of the sea, and which forms the gravel under better soil: What does it reveal upon careful inspection other than little stones or clear streams, like glass that has been ground into pieces through motion, through melting, or in some other way? And the action of fire can easily do this if salts are added, and in fact salts were not absent at the beginning.

[IV. *Sea salt, fires, and cycles of precipitation*]

From this previously observed genesis of things comes sea salt. For just as burnt materials attract moisture when they cool—whence oils are generated *per deliquium* in chemists' cellars—so will one easily believe that, in the beginning, before light was separated from darkness and while our globe still burned, the moisture, thrust out by fire, escaped into the winds; after that, just as in distillations, the regenerated moisture soon condensed back into watery vapors as the heat subsided; finally, as the cooled surface of the earth swallowed the moisture, it reverted to water, which washed the face of the earth, thereby drawing into itself the vast vestiges of the empyreuma[9] and the fixed salt. This produced a kind of lye, which then flowed into the sea. We certainly know, from the analysis of plants, that there are two kinds of fixed salt in lye: the one is, as the artisans say, an alkali, whose name comes from the plant that we know as soda, and which the Arabs call *kali;* the other is sea salt, which is more acidic. Last of all, it is plausible that the crust, shrinking as it cooled, left behind great bubbles proportional to its size, that is, hollows under huge vaults, which enclosed air and moisture, as happens with metals and other things that become more porous through melting. Then the crust separated into certain sheets and, according to differences of material and heat, came together unevenly in clumps; indeed, it would have ruptured in places, with the broken fragments tumbling into sloping valleys,

---

9. The burnt-smelling, tarry material that appears toward the end of a distillation at high temperature.

unde jam tum montes superfuere. Accessit pondus aquarum, ad alveum sibi parandum in molli adhuc fundo. Denique vel pondere materiae, vel erumpente spiritu, fracti fornices, maximaeque, humore cavitatibus per ruinas expulso, aut sponte montibus effluente, secutae inundationes, quae cum deinde rursus sedimenta deponerent per intervalla, atque his indurescentibus, redeunte mox simili causa, strata subinde diversa alia aliis imponerentur; novata est saepius facies adhuc teneri orbis.[12] Donec quiescentibus causis atque aequilibratis, consistentior emergeret status rerum. Unde jam duplex origo intelligitur firmorum corporum; una, cum ab ignis fusione refrigescerent, altera cum reconcrescerent ex solutione aquarum. Neque igitur putandum est lapides ex sola esse fusione. Id enim potissimum de prima tantum massa accipio ac basi terrae; nec dubito, postea materiam liquidam in superficie telluris procurrentem, quiete mox reddita, ex ramentis subactis ingentem materiae vim deposuisse, quorum alia varias terrae species formarunt, alia in saxa induruere, e quibus strata diversa sibi super imposita diversas praecipitationum vices atque intervalla testantur.

[v]

Haec vero utcunque cum plausu dici possint de incunabulis nostri orbis, seminaque contineant scientiae novae, quam Geographiam Naturalem appelles, tentare tamen potius, quam astruere audemus. Nam etsi faveant sacra monumenta,[13] tamen illis judicium deferimus, quibus interpretandi jus est. Et licet conspirent vestigia veteris mundi in praesenti facie rerum, tamen rectius omnia definient posteri, ubi curiositas mortalium eo processerit, ut per regiones procurrentia soli genera et strata describant. Neque vero omnem terrae scabritiem aut fundi naturam primae concretioni imputavero. Sufficit a generalibus causis duxisse sceleton ipsum, et velut ossamenta terrae exterioris, et totius structurae summam. Possunt

12. B: teneri adhuc orbis saepius novata est.
13. B: sacra divinorum oraculorum monumenta.

so that the harder parts, like columns, occupied the highest place. That is why there were mountains even then. Then came the weight of the water, seeking itself a bed in the soft ground. Finally, either the weight of the material or the bursting forth of the air smashed the vaults, so that the water in the hollows was pushed out through the ruins, or flowed down the mountains of its own accord; there followed the greatest floods, which, in turn, deposited sediments at different times. After these hardened, and with the return of similar conditions, different layers would have been placed on the first ones. The face of the then still tender globe often changed in this way, until, eventually, as these conditions subsided and came into equilibrium, there emerged a more settled state of things. This explains the twofold origin of solid bodies: first, that they cooled after being melted by fire; second, that they hardened again after being dissolved by water. Nor should one therefore suppose that stones arose through melting alone, though I certainly accept this as the basis of the earth and the principal part of its initial mass. I do not doubt that, later, as soon as the liquid material rushing over the earth's surface came back to rest, it deposited a huge quantity of matter in the pulverized debris. One part of it formed different kinds of earth, and another part hardened to stone, with various layers superimposed on one another, testifying to the different cycles and intervals of precipitation.

[v. *The many changes in our globe after its initial creation*]

This theory about the newborn globe may be plausible, and it may even contain the seeds of a new science called natural geography, but we venture to explore rather than to build. For even if the sacred monuments favor this reading, we nevertheless defer judgment to those who have the right to interpret. And even if the vestiges of the old world conform to the present appearance of things, our descendants will be able to explain everything better when human curiosity will have advanced far enough to describe the kinds and layers of earth that extend through the various territories. Nor do I want to attribute every roughness of the earth, or quality of the land, to that first hardening. It is enough to have deduced its skeleton — like the first bones of the outer earth — and its whole structure from the most general causes. For this really is possible, if you ask how the vast hollow of the ocean and the huge mass of the moun-

enim haec vera esse, si quaeras, unde immanis Oceani cavitas, et insanae montium moles sint natae, velut Imai continuatum Caucaso et Tauro jugum, Atlasque Africam protegens, et Lunae montes, quibus Aethiopia habitabilis facta est, et maximus editissimorum cacuminum per Americae longidutinem tractus, quo illa contra utrumque Oceanum obfirmata est, et in Europa nostra inconditi Scandinaviae scopuli, et Alpes a Pannonia pene ad Hispaniae fines Italiam praecingentes, et postremo noster in Saxonia excelsissimus Melibocus in eo tractu, quem a resinosis arboribus non obscuro Hercyniae vestigio, vocamus Harzicum nemus, cujus potissima pars Brunsvicensi ditione continetur. Sed non ideo negamus, solidato jam, ut nunc, globo, minores exustiones, et terrae motus, et privatas eluviones et sedimenta restagnantium aquarum supervenisse, quae magnos saepe tractus obtinerent converterentque; quorum apud nos quoque[14] vestigia mox dabuntur. Nec dubium est, freta alicubi perumpente mari effracta; terras in voraginem absorptas, et stagno mutatas; nunc inundata littora, nunc destituta; oppressa loca inferiora, et angustias ruinis montium interclusas, intercepto aquarum cursu; vicissim erumpentes via vi facta lacus, et excavatas ad effluxum valles; natos Vulcanios montes, et denatos; lateque sparsos pumices, et vestigia incendiorum impressa. Sed quid privatis imputandum sit, aut publicis causis, facilius aliquando statuet posteritas, explorata melius humani generis sede.

---

14. B: horum enim apud nos quoque.

tains came into being: the Himalayas, which extend through the Caucasus and Taurus ranges;[10] Atlas,[11] which protects Africa; the Mountains of the Moon, which made Ethiopia habitable;[12] the greatest extent of the highest peaks across the length of America, which fortify it against the ocean; in our Europe, the rough cliffs of Scandinavia and the Alps, which stretch almost from Hungary to the ends of Spain, surrounding Italy; and finally our own Melibocus, the highest mountain in Saxony, which lies in the mountain range that we name after resinous trees—whose still discernible traces are in the word Hercynia—which we call the Harz, and whose greatest part belongs to the House of Brunswick.[13] But we do not therefore deny that the globe, then already solid like today, experienced smaller fires, earthquakes, isolated floods, and deposits from floodwaters, which often occupied and changed large areas. We will detail the traces of these events, which one finds around us, below. There is no doubt that sea ruptures ripped open channels in certain places, that the land was torn deeply and transformed into a sea, that coasts were sometimes flooded and sometimes emptied of water, that lower-lying places were buried and channels were blocked by rubble from the mountains (so the flow of the waters was thus interrupted), that these waters then erupted onto a path fashioned by force, that volcanic mountains arose and passed away, that pumice stones were thrown far and wide, and that the traces of conflagrations impressed themselves upon the earth. But as to what can be attributed to special and what to general causes, posterity can determine that better after the seat of humanity has been investigated further.

10. The great alpine zone believed to traverse Asia from west to east was sometimes referred to as Taurus, Caucasus, or Imaus. The Imaus, or snow hills, correlate roughly with the Himalayas.

11. The Atlas chain of mountains in northern Africa, which extends from southwestern Morocco to northeastern Tunisia, forms a barrier between the coastal plain and the northwestern Sahara Desert. The main ridge of the chain, called the Great Atlas, exceeds ten thousand feet at its highest point.

12. Ptolemy wrote in 150 CE that the "Mountains of the Moon" provided the lakes of the Nile with snow water. We know them today as the Rwenzori Mountains of Uganda. See Ptolemy, *The Geography*, 4.8.

13. Leibniz's employers, the dukes of Brunswick-Lüneburg, were based in Hannover. Between 1635 and 1788, they shared revenues from the so-called "Communion Harz" with their relatives, the Brunswick-Wolfenbüttel line.

Quemadmodum autem omnia initio ignis corripuit, antequam lux a tene-
bris secessisset; ita restincto incendio omnia deinde aquis mersa censentur. Res sacris nostrorum monumentis[15] traditur; consentiunt antiquae
gentium narrationes, sed maxime[16] mediterranea maris vestigia adjuvant
fidem. Nam et cochleae in montibus peregrinantur, et ut nostra attin-
gam, succinum, quod in marinis legi solet, nonnunquam procul a pelago
et in nostris quoque oris effossum est. Et glossopetrae Melitensibus similes, hoc est canum marinorum dentes, prope Lüneburgum eruuntur; de
quibus omnibus mox dicemus distinctius. Sunt,[17] qui eo usque licentia
conjectandi procedant, ut tegente omnia oceano animalia, quae nunc terram habitant, aliquando aquatica fuisse arbitrentur, paulatimque, des-
tituente elemento, amphibia, postremo in posteritate sua primas sedes
dedidicisse. Sed pugnat ista[18] cum sacris Scriptoribus, a quibus disce-

dere religio est. Illud nunc dispiciendum, unde suppedi-
tata tanta moles aquarum, quae montes superaret, et quo
deinde delata, ut arida[19] sibi redderetur. Quidam ingenioso
magis, quam expedito commento rem sola mutatione centri terrae peragunt: ita gravium inclinationem alio versam,
et servata licet superficie, tamen altitudinem et humilita-
tem locorum permutatas, quod istae non per se sed centri
vicinia aestimentur, praesertim si quaedam centri vac-
cilatio[20] intelligatur in diversas partes, ita enim ab omni
latere erit vicissitudo elevationis et depressionis. Et sunt, qui magneticae
variationis experimentis adducti in tellure nostra aliud corpus ingens,

15. B: sacris religionis nostrae monumentis.

16. B: si ab his recesseris.

17. B: Equidem haut ignoro esse quosdam.

18. B: Sed praeterquam quod ista cum Sacris Scriptoribus, a quibus discedere reigio
est pugnent, hypothesis ipsa in spectata immensis difficultatibus laborat.

19. B: arida terra redderetur.

20. B: Quibus forte aliquis daretur locus, si maria et montes separatim in deversis
globi partibus consisterent, nec in eodem haemispherio permiscerentur. Quamquam
et hic intelligi possit quaedam centri vacillatio in diversas partes, ita enim ab omni
latere erit vicissitudo elevationis et depressionis.

[VI. *What was the source of the water that covered the earth? And where did it go?*]

As in the beginning, before the light had separated itself from darkness, fire seized everything, just so does one reckon that later, after the fire had been extinguished, everything was plunged under water. These things have been passed on through our sacred histories, which agree with the old stories of other peoples, but the inland vestiges of the sea offer the best support. For seashells have been transported onto the mountains, and (to touch upon our region) amber, which one usually finds near the seashore, is also sometimes mined far from the sea, within our own borders. Moreover, glossopetrae (that is, sharks' teeth), like those from Malta, have been unearthed near Lüneburg. We will speak about all of these things in greater detail later. There are those who take the freedom to conjecture so far that they have imagined how once, when the ocean covered everything, animals that now live on land were aquatic; then, as the water departed, these animals became amphibians, until their descendants eventually left that original home. But that conflicts with the sacred writers, with whom it is impious to disagree. We now have to investigate the source of so much water which rose above the mountains, and where it eventually flowed when they became dry again. Some, by means of a scheme more clever than it is clear, explain the matter purely through a shift of the earth's center;[14] according to this theory, the descent of heavy things changed direction and, though the surface was preserved, yet the height and depth of places changed completely; they cannot therefore be measured for themselves, but only according to their distance from the center, particularly if a certain vacillation is perceived in different parts, so that there will be alternating elevation and depression on all sides. And there are those who, misled by the experiments on magnetic variation, would believe that there is another great body inside our earth, like the kernel of a nut, which is not without its own motion. Now, if one takes such a body into consideration, doesn't one have to think as much about the attraction of weight at the center as at the magnetic poles? Which could lead one to believe that the body, having no fixed place, swayed back and forth. But there are easier ways to understand where the extra water went, and how the earth was freed of it. Namely, the water could have

14. Cf. Bernier [1684] 1992, vol. 2, 323–331.

tanquam nucleum in nuce, comminiscuntur, suo quodam motu non destitutum; quod si ergo in illo quaerendum, non magis respiciendum ad polos ejus pro magneticorum, quam ad centrum pro gravium attractione? Quod aliquando[21] cum corpore suo non satis adhuc certa sede nutasse credi posset. Facilius intelligitur, quorsum pervenerit aquae superfluum, ut terra exoneraretur. Potuit enim per caecos aditus tum primum diruptos recipi cavernis immanibus, et in globi interiora penetrare; cum quicquid aquarum olim montium praesentium cacumina tegere potuit, si miliaribus Germanicis quatuor maris aequore superiora credantur, nondum sexuagesimam faciat partem reliqui globi. Ita nihil prohiberet, quae nunc cernimus, emersisse, aut si pars altior jam ante exstabat, tum demum homines a summis jugis, ut Scythae objiciebant Aegyptiis, in novas sedes, ut celebratam Mosi vallem Sinear descendisse: forsan cogente etiam frigore, cum recessu maris ad inferiora ipsi quasi in altiorem aeris regionem sublati viderentur, quae jam non aeque vaporibus temperaretur. Sed si aqua firmata jam tellure, a depressiori loco in ipsos altissimos montes naturali causa ascendit in diluvio, alia adhuc molitio accersenda est. Pluviae per se non sufficiunt, nisi aer olim multo quam nunc aquasior fuit. Oceanum spiritu ex terra circum erumpente velut fornice sustentatum undique surrexisse in orbem parum credibile est: intumuisse velut inflatam ipsam telluris superficiem ad pristinum fusionis tempus pertinet, et Vulcani regnum, quo tempore massa mollis tenaxque erat; in duram jam et fragilem crustam non cadit. Externa, ut Cometae transitum in vicinia, aut Lunam propiorem, quam nunc, quibus attrahentibus aquae emicuerint, accersere non ausim. Neque mutatae gravium directioni centrove confido. Quodsi obviis insistendum est, nil propius videtur, quam ut credamus fracto telluris fornice, ubi infirmioribus fulcris sustentabatur, ingentem massam nudatis cacuminibus in subjectum anteaque inclusum mare procubuisse. Ita aquas antris expressas supra montes exundasse, donec reperto novo in Tartara aditu, refractisque repagulis claustrorum interioris adhuc terrae, quicquid nunc siccum cernitur denuo deseruere. Itaque si aqua telluris crustam semel inde a formatione texit, sufficit unus fornix; sin montes nova eluvione oppressit bis fornicata erat, exteriorque cavitas aqua, interior aere farta; ita priore rupto aqua in montes ascenderit, mox posteriore fracto in abyssum ulteriorem penetrarit, terrestribusque habitatoribus iterum indulserit in sicco locum.[22] In quibus sane

21. B: et hoc aliquando cum corpore suo.
22. B: locum veresimile est.

penetrated to the inner depths of the globe through hidden passages that were just then ripped open for the first time, before being swallowed up by vast caverns; for the amount of water necessary to cover the peaks of today's mountains—assuming that these lie four miles above sea level—does not even constitute the seventieth part of the rest of the globe. Nothing, therefore, precludes the possibility that what we see now emerged from the water, or that, if a higher section had already protruded, people came down from the highest mountain ridges, as the Scythians argued against the Egyptians, to settle in new places, like Moses's famous plain of Shinar.[15] Maybe they were also driven out by the cold, since, as the sea receded, it was as if they had been lifted into a higher part of the sky that was no longer tempered by the moisture. But if, after the earth had already hardened, the water rose in a flood from a lower place up to the highest mountains by natural causes, then we need to introduce yet another contrivance. The rain by itself does not suffice, unless the air was once much wetter than it is now. That the ocean rose everywhere across the globe, lifted as if on a vault by vapor that burst forth all over the earth, is hardly believable. The bulging of the earth's surface, likewise swollen, belongs to the initial time of the melting, during the reign of Vulcan, when the material was supple and pliant. It does not suit the then already hard and brittle crust. I do not dare to adduce accidents, such as the near passage of a comet, or a moon that was closer than today, whose attraction could have caused the water to gush forth. Nor do I trust a change in the direction or the center of heaviness. If, then, one is to take the direct path, nothing appears more sure than our belief that the vault of the earth collapsed at the point where it was buttressed by weaker supports, that a huge mass then crashed into the sea which lay under it and had previously been enclosed, and that the mountain peaks were thereby exposed. Having thus been forced up out of the caverns, the waters flooded the highest mountains until, with the discovery of a new entry to Tartarus and the destruction of the barriers to the previously closed interior of the earth, they withdrew once again from what we now perceive as dry. If, then, water covered the earth's crust once after its formation, a single vault suffices; but if water submerged the mountains with a new flood, then the earth's crust was double vaulted: the outer hollow filled with water, the inner one with air. So when the first vault ruptured, the water probably

15. Leibniz refers here to M. J. Justinianus, "Account of the Scythians and of Their Actions," 1.17, in *Philippic History*, bk. 2; for Shinar see Genesis 11:1–2.

explicandis juvare nos possunt aliquae ingeniosi Scriptoris meditationes, qui nuper sacram telluris theoriam dedit, montesque etiam et valles[23] ex ruinis formavit, scriptaque nonnulla eruditorum hominum, quorum ille studium excitavit. Nec abhorrentia quaedam de ruinis et sedimentis cogitaverat jam ante Stenonius, non contemnenda Europae parte lustrata, et fractorum fornicum vestigiis passim notatis, ut saepe ipsum nobis narrantem audire memini, ac gratulantem sibi, quod sacrae historiae et generalis diluvii fidem naturalibus argumentis, non sine pietatis fructu, astrueret. Sed nobis altius iri visum est, fornices ex fusione, maria deliquio salium vapores aqueos resorbentium esse formata.

[VII]

In nostro tractu nihil est extantius Meliboco monte; hujus cacumen maxima anni parte inaccessum et strigium choreis apud credulos infamatum. Accolae *Bructerum* vocant, vulgo *Brocken;* non a Bructeris, populis apud verteres memoratis, sed ab eadem tamen causa, quae his nomen dedit. *Broeck* enim Saxonibus terra est humida, et in paludem vergens: Quale solum hujus montis. Et quominus dubites, substat majori Bructerus minor (*Kleine Brocken*); pari conditione terrae. Duae autem causae paludosum fundum facere solent: una manifesta aspectui, cum aquae decursum, nisi lentum, non habent; altera nostris, ut memini, fossoribus metallorum notata; ipsa scilicet (quod minus rere) firmitas soli subjecti, cum sub limo superficiario occultatae rupes aquis percolaturis ad interiora aditum negant, utili indicio opera molientibus. Summum Bructerum ob nives non nisi media aestate ascendas. Christianus Ludovicus Dux[24] muniri curaverat viam. Supra rivulus occurrit, et scio esse, qui argumento fontium in celsissimis montibus nascentium refellere sperant originem fluminum ex coelesti aqua; quasi praeter ipsum naturale fluentium pondus alius mo-

23. B: valles ex ruinis formavit.
24. B: Serenissimus Dux Christianus Ludovicus.

moved up the mountains; next, after the destruction of the second vault, it might have penetrated into the abyss farther below, thereby affording the denizens of the earth a dry place again. For a sound explanation of these things, we can turn to some thoughts from the clever writer who recently offered a *Sacred Theory of the Earth*,[16] and who would construct mountains and valleys out of collapses, and to the works of several other scholars whose diligence he inspired. Not averse to such things, Steno had already thought this way before about collapses and sediments, after visiting a considerable part of Europe and noting the vestiges of broken domes in various places. I remember hearing him tell us about this often, and that he rejoiced in contributing, through natural arguments, and not without benefit for piety, to a belief in the sacred history and the universal flood. But, going further, I argue that the vaults were formed through fusion, while the seas were formed when salts reabsorbed watery vapors through deliquescence.

[VII. *Bructerus and the origin of springs*]

In our region, nothing is higher than Mount Melibocus. Its peak, which is inaccessible for most of the year, is notorious among the credulous for witch dances. The locals call it Bructerus or, commonly, Brocken. This name does not stem from the Bructerites, a people well known among the ancients, but rather from the same source to which they owe their name: for the Saxons, *Broeck* is a damp earth tending toward the swampy, just like the ground on this mountain. And to remove any doubts, a smaller Brocken (Kleine Brocken), with the same kind of earth, lies beneath the bigger Bructerus. Two causes tend to make the ground swampy: the first is obvious—that is, the waters have nothing but slow drainage; the other reason, as I recall, was observed by our miners and consists in (what is less obvious) the firmness of the underlying ground, when stones hidden beneath the shallow mud keep the percolating water from reaching the interior, a useful sign for those toiling at work. Duke Christian Ludwig provided for the construction of a path. At the top is a little stream, and I know there are those who, by reason of springs that originate on the highest mountains, hope to reject rain as the source of rivers, as though

16. Burnet 1681, bk. 1, chap. 12.

tor salientibus interveniret; ut in animalibus sanguis calore elevatur, et artificiis aqua in jactus exprimitur. Sed rivulus Bructeri non in ipso apice nascitur, nec nisi de superiore adhuc loco exonerat superfluum humentis terrae: idemque alibi contingere vix ambigo. Late autem prospicitur mons, et prospicit in Germanicum mare. Nec id mirum accidere potest Geometricas rationes accersenti. Nam in eadem regione Hercyniae, sed loco multo inferiore, Cellerfeldae, oppidi metallariis habitati, incendium ante aliquot annos ex Hamburgi propugnaculis noctu visum accepimus.

[VIII]

Metalla autem non in ipso Bructero, licet mineralium rerum haud experte, sed mediocribus Hercyniae jugis, aliqua et sub pedibus montium fodiuntur; tantaque eorum copia est, quantam alibi vix hodie monstres uno loco nostri continentis. Nam vel plumbi nigri solius intra leucae Germanicae spatium quotannis incredibilis copia ex venae materia eliquatur. Vena est velut folium quoddam, sive stratum, sub terra longe lateque pro-

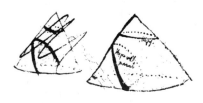

currens, crassitiei mediocris, in quo peculiare genus terrae, saxi aut metalli ab ambientibus divisum; nec melius quam conicarum sectionum similitudine illustratur. Nam quae nostris venae pendentes, *schwebende Gänge,* dicuntur, eae terminantur in circuli aut Ellipseos modum; quae vero cadentes, *fallende Gänge,* eae quasi in infinitum descendunt ad inferiora, ut Hyperbolae parabolaeve Geometris notae. Et propria cuique generi jura sunt in portionibus assignandis, propriae fodiendi rationes. Unde putei cadentium venarum jus profunditatis habent in infinitum, caeterarum nihil minus. Et qui pendentes angustiores exercent fodiendo, ut nuper Osterodam Islebia vocati, torticollarum nomen, *Krummhälsse* accepere ab operis modo, cum erecti esse non possint. Jam credibile est, multa strata, quae olim horizontalia erant, cum fornices telluris in sua integritate perstabant, postea subsidente crusta, ruinisve orbis, inclinata fuisse in de-

some motor other than the natural weight of the flowing water—like blood lifted through heat in animals, or propelled water forced upward with machines—had to be at work in springs.[17] But the little stream on Bructerus does not originate on the peak itself. Rather, it only unloads the spillover from the damp earth, which itself comes from an even higher place. I hardly doubt that the same thing also happens in other places. The mountain is visible from far away, and from it one can see to the German Sea.[18] This is no wonder, if one employs geometric measurements. For we heard that, some years ago in the same Harz region, but in a much lower place called Zellerfeld—a town inhabited by miners—one saw a night fire from the outskirts of Hamburg.

[VIII. *Deposits of metal in the earth and a description of veins*]

But metals are not excavated on Bructerus itself, which is not lacking in minerals, but rather on the intermediate ridges of the Harz, and also at the foot of the mountains, in quantities so great that one hardly finds such in another place on our continent today. For within the space of one German mile,[19] the vein lodes yield an incredible amount of black lead ore alone each year. A vein is like a sheet or layer that stretches far under the earth, is of middling thickness, and contains a special kind of earth, stone, or metal that is separated from its surroundings. It cannot be illustrated any better than in its similarity to a conic section. For what we call hanging veins, or *schwebende Gänge*, are bounded like an ellipse or a circle; but what we call falling veins, or *fallende Gänge*, descend into the depths as if to infinity, like the hyperbolas and parabolas known to the geometers.[20] And each kind has proper rights, which are allotted in shares according to the special accounting of the mines. So, with falling veins, the shafts have the right to proceed to infinity, which is not at all the case with the other kind. And the miners who work the narrower

17. Cf. Descartes [1664] 1983, 213–214 (IV, §64–65).

18. The North Sea.

19. A German mile was equivalent to between four and five English miles.

20. The A manuscript contains sketches in which Leibniz illustrates the point (to himself at least) with conic sections. We have reproduced these with the Latin text.

scensum. Quid enim naturalius, quam in ipsa formatione ex liquido, sive cum terrarum orbis primum concrevit, sive cum inundationes magnae postea sedimenta deponebant, suum quaeqe locum a pondere cepisse, et ad libellam composita horizontis planum affectasse lege fluidorum, cujus aequabilitatem deinde vis major concussis fundamentis deformavit. Et ratio per se valida praesenti oculorum testimonio intenditur. Audio in Norvegiae promontoriis ex praeruptis et velut ferro abscissis passim immensis rupibus prodeuntes in mare notari strias, foliorum ingentium terminatrices. Passim etiam valles aquarum vi, aut alio impetu effractae vel excavatae, ab utroque latere oppositos parietes montium, simili stratorum genere variegatos ostentant. Meminique cum non ita pridem Osterodae in Hercyniis pendens vena ardosiae aeriferae ferro aperta esset, continuationem in opposito vallis latere notatam. Videmus et concursu ditescere aut dilatari, contra divaricatione minui venas, et in centro plurium quasi nodum quendam metalli cumulati intumescere (*Stock*) quo vastum aliquando spatium occupatur, ut in Goslario monte Rammelo. Saepe etiam notatum est, post ditioris metalli nidum sterilescere venam, quasi vicini tractus opes olim vi ignis aut humoris in unum confluxissent, ut in monte S. Andreae. Persaepe etiam subterranei metallorum procursus vallibus et rivis sub dio respondent, ut primaria superiorum Hercyniae fodinarum linea rivo Cellae. Ubi conjicias telluris rimas supra in vallem dilatatas intus venarum cadentium forma descendisse, metallo deinde aut saxo, aut terrae genere aliquo, sive vi ignis colliquantis, sive aquarum affluxu oppletas. Certe rubricam fabrilem alicubi manifeste notavimus inter schisti lapidis hiatus ab aquis insinuatam; nec aliter boli colliguntur. Non tamen facile simplex jam solidi ruptio venam fecit. Sed plerumque inter indurescendum cum crusta haec telluris formaretur, separationis lineamenta coepere; nam venae utroque latere in fibras abire solent, et fibrae in minimas saxi commissuras disperguntur, prorsus ut in animalibus vel plantis vasa majora in capillaria filamenta discedunt, tandemque in stamina oculis imperceptibilia evanescunt. Nec sane dubito, diligenti observatione principia constitui posse aliquando, unde nascantur regulae conjiciendi de abditis sub terra, velut foliis metallorum, longe illis meliores, quae passim ex levi causa, et praesertim ex mundi plagis receptae inter fossores, magis traditione quam successu celebrantur. Nam quae genera in ipsa terrae superficie exeunt in lucem (*zu Tage ausstreichen*) aut in fodina innotuere, ea ex stratorum procurrentium legibus indicia in viciniam extendent. Et peculiares sibi matrices unum quidque genus amat. Ita galena in spatho frequens habitat, specularis lapis saepe Alabastritae

hanging veins were called the crooked necks, or *Krummhälse*, because they could not stand up straight owing to their manner of working, like those recently summoned from Eisleben to Osterode. Now it is likely that many layers, which once lay horizontal when the vaults of the earth still stood in their entirety, sank toward the deep along with the collapsing shell and ruins of the globe. For what is more natural than that each layer, as it formed from a fluid—either when the earth first hardened or later, when great floods deposited sediments—assumed its place according to weight or was deposited according to water level? And what could be more natural than that each layer sought, in keeping with the laws of fluids, to form a horizontal plane whose regularity was later destroyed by a powerful force that disrupted the foundations? This reasoning, itself already valid, is strengthened by the testimony of the eyes. I hear tell that in the promontories of Norway there are places where striations stretch into the sea from huge, scattered cliffs that have been broken apart and smashed off, as if with a hammer; these are the ends of enormous sheets. Sometimes valleys, broken open or hollowed out by the force of the water or some other violent impulse, reveal facing mountain walls with similar kinds of variegated layers. And I also recall that, not long ago, a hanging vein of copper-bearing schist (*Kupferschiefer*) was discovered in the Harz at Osterode, whose continuation was found on the opposite side of the valley. We see that these veins become richer and wider as they run together, or smaller as they diverge, and that a knot of accumulated metal (*Stock*) swells up in the center of most. Sometimes, as in Rammelsberg near Goslar, this knot occupies a considerable space. It has often been observed that a vein turns barren after passing through a rich metal nest, as if the riches of the surrounding area had once, through the force of fire or moisture, all flowed into a single place, as at St. Andreasberg.[21] Very often the subterranean ore veins coincide with the valleys and rivers aboveground, as, for example, the main line of the most important Harz mines with the Zellbach.[22] One can therefore assume that crevices, which opened into valleys above, stretched out like falling veins below.

21. Rich silver ores were discovered at Andreasberg in the Harz at the end of the fifteenth century. As Leibniz suggests here, Andreasberg was as famous for its dramatic bust, when rich deposits dried up, as it was for the great discoveries of the early years.

22. The Zellbach runs directly between the Harz mining towns of Clausthal and Zellerfeld, the administrative mining centers for the House of Brunswick-Lüneberg and the House of Brunswick-Wolfenbüttel, respectively.

conterminus jacet, pyrites cum aere et sulphur praebet, schistum quoque ardosium aes diligit, argentum vivum in cinnabari latet. Prodiit ante annos non ita multos quidam Alexandri Achillis nomine sumto, homo militaris et peregrinator: is quod votis magis, quam spe praecipimus, delineare ausus est magnae parentis ossa et procurrentes per corpus ejus venas; quasi retectam tellurem perspexisset, perlustrassetque unus mortalium,

Omnia sub magna labentia flumina terra.

Sed tamen vel cogitasse laudandus est, quod longo saeculorum studio quaeri mereatur. Metallurgi passim vulgari notione venas pro truncis ramisque habent, quasi vegetatione crevissent: scilicet quia delineatas a mensoribus hanc speciem aliquando praebere vident. Nec dubium est, cum prima telluris tenerae stamina duceret conditor, aliquid formationi animalis aut plantae simile contigisse, sed incendiis et eluvionibus et ruinis nunc ita detortum perturbatumque in hac superficie, et velut cute, ut aegerrime nosci possit.

These then filled up with water, ore, stone, or some other kind of earth through the melting power of fire or flowing water. Certainly we once observed red ocher that had clearly been washed into the gaps between the schist rocks. One collects bolus stones in no other way. Still, a vein is not merely a straightforward, simple rupture in something otherwise uniform. Rather, distinct lines arose during the hardening process, as the earth's crust was forming. For the veins generally dwindle into fibers, and these fibers scatter into the smallest rock seams, just as the larger vessels in animals and plants divide into hairlike filaments and then finally disappear into threads that are invisible to the eye. And I do not doubt that someday, through careful observations, it will be possible to construct principles that can yield rules governing what is hidden under the earth, like metal ores; and I believe these rules can be far better than those generally praised on weak grounds, more through tradition than success, and above all because the miners of some regions apply them. For those kinds reaching up to the earth's surface (*zu Tage ausstreichen*), or those that are observed in the mine, will leave signs throughout an area, according to the laws of projecting layers. And every kind loves its special mother. Thus, galena frequently abides in spar, specular stone[23] often lies near alabaster, pyrites promise both copper and sulfur, schist also likes copper ore, and quicksilver is hidden in cinnabar. Not many years ago a certain soldier and traveler appeared, who called himself Alexander Achilles.[24] He dared to draw the bones of the great mother and the veins running through her body, as if he had inspected the naked earth and, as the only one among mortals, had scrutinized

"all streams beneath the mighty earth that glide." [25]

But he ought nevertheless to be praised for speculating about something whose acquisition merited many centuries of study. Miners generally believe, in common fashion, that veins are stems and branches, as if they had grown like plants—evidently because they see how these veins sometimes look like that in the drawings of subterranean geometers. There is no doubt that something like the formation of plant or animal occurred when the creator wove the first fabric of the tender earth. But this got so confused and distorted by fires and floods and collapses in the earth's surface, which is like a skin, that it became very difficult to recognize.

23. Species of gypsum, mica, selenite, or talc.
24. Prussian nobleman, died 1675.
25. Virgil, *Georgics*, 4.366.

Operae pretium autem facturum arbitror, qui naturae effecta ex subterraneis eruta diligentius conferet cum foetibus laboratoriorum, sic enim
Chemicorum officinas vocamus, quando mira persaepe in natis et factis
similitudo apparet. Nam etsi ejusdem rei plures causas in potestate ha-
beat ditissimus rerum parens, amat tamen et constantiam in varietate. Et
magnum est ad res noscendas vel unam producendi rationem obtinuisse:
quemadmodum Geometrae ex uno modo describendi figuram omnes ejus
proprietates derivant. Praeterea ubi similia instrumenta et vasa occurrunt,
ignis cum sulphuribus, aquae cum salibus, et genera terrarum lapidumve,
communia et nostris et naturae officinis, tutius similem cognatis corporibus originem assignes, quam diversam nullo experimento cognitam ex ingenio fingas. Neque enim aliud est natura, quam ars quaedam magna: nec
semper toto genere a nativis factitia distiguuntur; nec refert eandemne
rem Daedalus aliquis Vulcanius in furno inveniat, an lapicida ex terrae
visceribus proferat in lucem. Et quanquam de nova per artem genera-
tione metallorum aut corporum similarium simpliciorum nihil affirmare
velim, nec satis dicere ausim, an aurum argentumque aut hydrargyrum,
vel salem etiam de novo produxerit, aut etiam vere destruxerit quisquam,
puto tamen, non minus etiam rarum esse ipsammet naturam deprehen-
dere edentem hos foetus: plerumque enim dudum alibi conceptos colligit
tantum detegitque. Vereor enim, ut fidem tueantur, quae narrant aliqui
de nascente iterum auro in expositis ad solem arenis jam tum elotis; aut
de rejectamentis minerarum vel laminis tegularum ipsa temporis longitudine ditatis. Temere fiditur narrationibus hominum, qui credulitate sua
alienave fallere fallive morem fecerunt; qui homunculos aut monachos
plutonios laboris subterranei socios vident: qui virgula divina abditos
tellure thesauros ruspantur, et tamen ubi oculos obligaveris, nec maxi-
mas notissimasque venas micantis baculi indicio agnoscunt. Auri semina
Corbachii Waldecciorum, aut ferri in Elva insula nequicquam quaeras.
Plinianae autoritati Stenonii diligentiam oppono. Neque uspiam compertum est nostris fossoribus excisum spathum cum plumbo suo rursus
vegetasse; non magis quam marmor erutum lapidicinis, aliave id genus,
quae sapienter Iureconsultus Romanus in usufructu esse negavit, cum
non renascantur. Non tamen nego puteos esse aut cuniculos, ubi denuo
agnascitur minera, et viae portaeque arctantur, quod in Rammelo prope

[IX. *The generation of minerals explained through chemistry*]

I believe that whoever more carefully compares the productions of nature with the fruits of the laboratories—that is indeed what we call the workshops of the chemists—will collect a reward, because an amazing similarity between natural and artificial things is often evident. For even if that wealthiest author of things has many causes for the same thing at his disposal, he nevertheless loves permanence within multiplicity. And it is a major step into the knowledge of things when one has established a single method for producing them. In this way the geometers derive all the properties of a figure from a single method of describing it. Further, where similar instruments and vessels occur, as fire with sulfurs, waters with salts, and the varieties of earth and stone—all common to both our workshops and to those of nature—you might more safely attribute similar origins to kindred bodies than if you cleverly invented a different account, one not known through any experiment. For nature is nothing other than a great art. And the entire class of artificial things is not always distinct from natural productions; for it is all the same whether some Vulcanius Daedalus discovers a thing in his furnace, or whether a stone-cutter brings it to light from the bowels of the earth. And I do not want to maintain anything about the artificial generation of metals de novo, or about similar simple bodies. Neither do I wager to say whether some gold, silver, quicksilver, or salt was ever produced de novo, or whether anything was ever really destroyed. Nevertheless, it is no less unusual to catch nature in the act of producing her fruit, for she mostly only gathers and reveals what has already been conceived elsewhere long ago. For I fear one cannot trust those who report that gold is reborn after sand that has previously been washed is exposed to the sun, or that the slags from mines or the plates of roof tiles were enriched over time. One blindly believes the stories of people who have accustomed themselves, through their own credulity and that of others, to deceiving and being deceived—people who see homunculi and Plutonic monks nearby while they work underground, and who search the earth for hidden treasures with divining rods, even though, when you bind up their eyes, they do not find the largest and most famous veins through the sign of the quivering rod.[26]

26. Cf. Agricola [1556] 1912, 38–41.

Goslariam temporis lapsu contingere scimus: Sed ibi aquae chalcanthum et mistam metallo materiam vehentes sedimentum deponunt, nec gignunt aes plumbumve, sed afferunt. Itaque mineras, id est varias metalli larvas quotidie produci video, sive ab arte sive a natura, sed de ipsis metallis nihil affirmo, neque arcanis rerum temere praejudicium interpono. Resuscitationes primi corporis facilius agnosco, et scio simplicia illa genera non aegre minus destrui quam produci: usque adeo facile reviviscunt. Satis constat, nullis Chemicorum torturis Mercurio confessionem mortis expressam. Et cum nitri Spiritus ex alcalico sibi corpus resumit, fortasse verum, sed tenue nitrum igni superstes novam habitationem incrustat, et cum regenerari dicitur, recolligitur tantum. Quemadmodum credibilius judico, sepultum in aceto liquorem ardentem ex saccaro Saturni resurgere, dum salibus a plumbo absorptis vis coercendi spiritus retinendique retunditur, quam spiritum vini veterem interire, et novum velut miraculo de metallico corpore nasci.

You will search in vain for seeds of gold in Korbach or Waldeck, or for seeds of iron on the island of Elba. I weigh Steno's careful observation against the authority of Pliny: nowhere have our diggers observed that hewn spar generated new lead, no more than did the marble excavated from quarries or other things of this kind, which, because they would not grow again, were quite reasonably excluded from usufruct by Roman jurists. I will not, however, deny that there are shafts and tunnels in which a mineral forms anew, so that the paths and entrances become narrowed, which we know is happening over time in Rammelsberg near Goslar.[27] But the waters there, which carry along copper vitriol and material mixed with metal, are depositing sediment; they are merely transporting the copper or lead, not generating them. Minerals, then, are the various masks of metal that I see produced every day, whether through art or nature; about metals themselves, however, I affirm nothing, not wanting to meddle in arcane things with blind prejudice. I more readily acknowledge the resuscitation of original bodies, and I know that simple kinds are no less hard to destroy than they are to create: they always recover so easily. It is well enough known that no torture by chemists has yet coerced mercury into a death confession. And if the spirit of niter takes up its body again from out of an alkali, then maybe a true but tender niter, having survived the fire, can build the crust of a new home; and when one says that it is reconstituting itself, it is actually just being reassembled.[28] Similarly, I judge it more likely that the flammable liquid buried in vinegar reemerges out of the sugar of lead when, by means of the salts absorbed by the lead, the vinegar's power of confining and retaining spirits is weakened, rather than that the old spirit of wine perishes and then, as if by a miracle, is born anew from a metallic body.[29]

27. The silver mines of Rammelsberg were among the oldest and most famous in all of Europe. With slag heaps dating back to the ninth century, and water tunnels dating from the twelfth, Rammelsberg was already legendary in Leibniz's time as the mountain that had bankrolled the first Holy Roman emperors.

28. This refers to the regeneration of saltpeter from the spirit of niter and fixed alkali, as reported by Robert Boyle in his "Essay on Nitre." See Robert Boyle, "A Physico-Chymical Essay, containing an Experiment with some Considerations touching the differing Parts and Redintegration of Salt-Petre," in Boyle 1669.

29. Here Leibniz refers to the fact that if lead calx (lead oxide) is dissolved in vinegar, filtered, and the solution evaporated, a substance called sugar of lead is produced. When this salt is destructively distilled in a retort, acetone is produced. The editors would like to thank Lawrence Principe for his help with this section.

Cum ergo plerumque res magis larvas sumant, quam naturam deponant, minus mirum est, tam multa laboratoriis et fodinis communia prodire. Nostro instituto suffecerit aliqua in specimen attulisse. Cinannabarin nativam Hydrargyri venae praebent, nec id in Idria tantum compertum prope Adriaticum mare, sed et in nostris oris juxta Walkenredam, ut produnt Annales, et nunc quoque reliquiae leguntur, quales et a Bructero a se afferi, sunt qui jactant. Nemo autem ignorat, Cinnabarin etiam ex sulphure et hydargyro per artem parari, ut de antimoniali illa nil dicam; nam et antimonio verum sulphur non deest, etiam arte eliciendum: magno indicio, quod nos in officinis facimus, naturam suo quodam modo in ipsis terrae recessibus egisse. Sublimatione enim utriusque parentis per vim caloris constat pulchellam sobolem nobis nasci. In vicinis Goslariae fornacibus Langesheimensibus, quas Brunsvicenses magistri exercent ad Indistrae flumen, eliquatur venae genus, quod plumbum atque aes praebet. Illic duo insignia naturae imitamenta visuntur: Zincum et minera lapidis, quem vocant calaminarem. Zincum fumo evectum intus adhaeret, moxque eximitur. Sed alius fumus parietem fornacis lentius incrustat, cujus fragmina dudum abjecta nunc pro lapide calaminari adhibentur, et aeri colorem aureum tribuentes faciunt, quod orichalcum appellamus. Interim aiunt Zincum nativum ex remotissimo Oriente a Batavis afferri. Et calaminarem lapidem etiam in Germania effodi constat. Et Amiantho similem materiam igne indomitam ex Pyrita Goslariae tertium usto exsudasse in summo, jam Georgius Agricola notavit. Unde prona suspicio est, quod exiguis speciminibus nos ludimus, naturam magnis operibus executam; cui montes sunt pro alembicis, Vulcani pro furnis. Auripigmentum quod in Turcica ditione copiose eruitur, alii arte tentant, et rubram sandaracam ex proprii generis minera educendam etiam sulphure et arsenico, id est cobalti fumo imitari licet; bismuthum autem nihil aliud quam ejusdem cobalti regulum seu corpus esse scimus, fortasse et a natura alicubi eliquatum.

[x. *Products common to laboratories and mines*]

Since, then, things generally assume disguises rather than changing their nature, it is not so surprising that laboratories and mines produce many similar things. For our purpose, it is enough to offer some examples. Veins of native cinnabar yield quicksilver. This has been observed not only in Idria,[30] near the Adriatic Sea, but also in our own region, near Walkenried (as the histories relate), where one can gather bits and pieces even today; there are even those who claim to have brought such things down from Bructerus. But everyone knows that cinnabar is prepared artificially from sulfur and quicksilver, not to mention cinnabar of antimony, since antimony does not lack true sulfur, which can be drawn out of it through art; this is an important sign that nature has already produced, in the hidden recesses of the earth and in her own way, what we make in the laboratories. For it is common knowledge that the sublimation of both parents through the force of heat yields the most beautiful offspring. In the furnaces of Langelsheim near Goslar, operated secretly by the masters of Brunswick, a kind of vein is smelted that produces lead and copper. Two extraordinary imitations of nature can be seen there: zinc, and a mineral called calamine.[31] Zinc is carried up with the smoke, adheres to the inside of the furnace, and is eventually removed. But there is another kind of smoke that coats the walls of the furnace more slowly, and whose pieces, discarded in former times, are now used as calamine stones. Insofar as they yield a golden yellow-colored copper, these stones make what we call brass. Yet, they claim that zinc was carried from the farthest reaches of the Orient to Holland. Certainly, calamine stone is also mined in Germany. The pyrites of Goslar, when burned for the third time, secrete a fire-resistant material similar to amianthus[32] onto their surface, as was noted by Agricola.[33] One is thus inclined to suspect that nature, using volcanoes as furnaces and mountains as alembics, has

30. The mines of Idria, about thirty miles from Trieste, provided a rich source of mercury. In Leibniz's time, they were controlled by the Austrian Habsburgs; today, Idrija (the name derives from the same root as hydrargyrus, the Latin for mercury) lies inside the borders of Slovenia.

31. Calamine was an important ore of zinc in England.

32. A kind of asbestos.

33. Cf. Agricola [1556] 1912, bk. 6.

[XI]

Ex auripigmento vi ignis attollitur Rubini quoddam seu carbunculi genus; fragilius quidem nativo, sed nos nec tam magnos adhibere ignes pos-
sumus, nec tam diuturnos, quam Vulcanus hypogaeus, ubi fortasse fusa
aliquando vix seculorum refrigeratione induruere: constat autem facilius
frangi, quae promtius constitere subita extinctione caloris. Unde nescio,
an omnis gemmarum origo ab aqua sit, ut vulgo sibi persuadent, certe
qui quaedam rudimenta gemmarum, fluores appellarunt (qualis ille est
caeruleus, ex quo phosphorus smaragdinus paratur, qui hausto tantum
calore lucet), fusionem in mente habuere. Verissimum esse fateor, materiam solutam mox cristallifico frigore figuras angulosque accipere, sunt
tamen quae non tantum aqua, sed igne solvuntur, nec tantum ex liquore,
sed et ex fumo in corpus recollecta geometrico naturae artificio figu-
rantur. Cristalli montanae, aliaeque gemmae, ipsique adamantes saepe
reperiuntur in cavitatibus saxorum, et in Drusis, ut vocant fossores, de
quibus mox aliquid dicemus. Apud Golcondam adamantes egregios aiunt
a sabulo praeberi, sed fieri potest, ut vi aquarum confricatuque glareae
diuturno et violento evolverentur gemmae putamine, in quo natae erant, cujus natura mollior non aeque duravit. Est igitur ubi nudae, est ubi
saxo inclusae deprehenduntur, ut verissimum sit, quod proverbio vulgi
nostri jactatur, ab ignaro lapidem in vaccam jactari, qui sit pluris, quam
vacca. Utrum autem ignis, an aqua formaverit, nihil generale adhuc constitui potest. Utrumque enim modum ars imitatur. Quanquam fortasse
spectabiles intra crystallum montanam animalculorum aut graminis
formae, et decurrentes guttae bullaeve generationi ex liquoribus favent.
Neque ego abnuerim, ad eum modum, quo alumen et vitriolum in vase,
postquam pars humoris calore fugata est, figuras accipiunt, multa etiam
saxeae durietiei corpora tunc nata videri, cum a magna materiae liquentis

accomplished in her mighty works what we play at with our little examples. Others have attempted to produce orpiment,[34] which is abundantly extracted in the Turkish empire, through other means. And red sandarach,[35] which one draws from a certain kind of mineral, can be imitated with sulfur and arsenic, that is, a fume of cobalt. We know, however, that bismuth is nothing other than the regulus, or body, of that same cobalt, and that it was perhaps smelted by nature herself in some places.

[XI. *The generation of precious stones, natural and artificial*]

A kind of ruby or carbuncle is created out of orpiment and with the force of fire. It is, to be sure, more brittle than the native kind, but of course we cannot use such great or lasting fires as the underground Vulcan, where something, once it is melted, may not harden through cooling for centuries. On the other hand, it is certain that whatever has hardened through the sudden extinction of heat breaks more easily. I do not know, therefore, whether all gems have their origins in water, as is generally believed. Doubtless, those who called certain simple gems fluorspar had melting in mind (of this kind, for example, is blue fluorspar, out of which emerald phosphorus is prepared; and the latter glows as it absorbs heat). I consider it absolutely true that dissolved material eventually develops angles and shapes owing to the crystallizing action of the cold. Of course, some materials are dissolved not only by water, but also by fire; others are gathered into a body not only from liquid but also from smoke and then formed by the geometric art of nature. One often finds rock crystals, other gems, and even diamonds in the rocky chambers and *Drusen*,[36] as the miners call them, about which we want to say something later. The sands around Golconda are supposed to yield excellent diamonds.[37] But it could be that these gems were torn out of the envelope in which they were formed—an envelope whose softer nature was not equally durable—by the power of the water, and by the violent and extended chafing of the

34. Orpiment (auripigment), known also as yellow arsenic, occurs naturally in soft, gold-colored masses. See glossary.

35. Red sandarach, or realgar, a naturally occurring disulphide of arsenic, commonly known as red arsenic, or red orpiment, was used as a pigment. See glossary.

36. Geodes.

37. Tavernier 1768.

solutione ad firmitatem natura imminuto humore vel aestu regrederetur. Interim et de talco ambigere cogor, quod crystallo affine est, et totum ex foliis consistit; jam vero et igne chemico praestatur aliquid simile sine crystallismo, cum terra tartari foliata sulphure inde sublimato paratur, tametsi haec folia ad naturalium durabilitatem non accedant.

[XII]

Habere naturam sublimationes suas non minus quam artem, dubitari non potest. De Ammoniaco certe jure dixeris, volatilem esse mineralis regni salem, sulphuri socium, unde et subterraneo calore eliciatur. Sane passim sulphureae naturae rivi gravem et quasi urinosum Ammoniaci odorem spirant, et prope Neapolim Campaniae vidimus in phlegraeo quodam campo ex foraminibus terrae exhalantem operculis capi. Et suspicamur similem ei originem esse, qui ex Oriente dicitur allatus, prostatque in officinis. Sane qui mittunt populi, rudes hodie, nihil artificiosae compositionis promittunt, qualem vulgo narrant, ex lotio Camelorum in arena cum sale marino, et (ut aliqui addunt) fuligine mista; quasi coeuntibus tribus naturae regnis, suspecto mihi artificio. Eundem tamen salem arte parare quidam spondent, et amicus meus[25] in fornacibus quibusdam casu notavit.

25. B: et amicus.

gravel. There are thus some places where one finds gems in the open, and others where one finds them enclosed in stone, so that our common proverb is very true: the stone that the fool throws at a cow is worth more than the cow.[38] But whether the fire or the water formed gems cannot be generally determined yet. For art imitates each method. Regardless, the shapes of little animals or grasses, and the running drops and bubbles that one can see in rock crystal, perhaps favor a formation from liquids. I will also not deny that many bodies that are now hard as stone seem to have arisen in the same way that alum and vitriol in a vessel take on shapes after part of the dampness has been chased away by heat: that is, nature, through a lessening of moisture or through heat, returned to hardness through a great dissolution of liquid material. For all that, I have my doubts about talc, which is related to a crystal and consists completely of leaves. In fact, something similar is obtained through chemical fire and without crystallization when one prepares foliated earth of tartar by subliming sulfur from it, even if these leaves do not attain a natural hardness.

[XII. *Natural sublimations and the preparation of sal ammoniac*]

There can be no doubt that nature, no less than art, has her sublimations. One can certainly say of sal ammoniac, and with justice, that it is a volatile salt of the mineral kingdom and a relative of sulfur, for which reason it is expelled by subterranean heat. In fact, streams of a sulfureous nature give off a strong urinous odor similar to sal ammoniac in some places. In Campania, near Naples, on the so-called Phlegraean Fields, we saw how sal ammoniac, breathing out of the earth's hollows, was captured with lids. And we suspect that the sal ammoniac offered for sale in the chemical workshops there, which supposedly comes from the Orient, is of a similar origin. At least, we cannot expect any such skillful manufacture from those now uncultivated peoples who send the sal ammoniac. The one usually reported — camel urine in sand with sea salt and, as some add, ashes — as if the three kingdoms of nature had come together, seems to me a suspicious piece of art. Nevertheless, there are those who claim to prepare the salt artificially, and a friend of mine happened to notice it in certain furnaces.

38. Popular proverb from the Upper Palatinate.

[XIII]

De argento aurove, aut alio metallo, quod statim suum est vel certe Obryzo accedit,[26] suspicari pronissimum est, non sine vi ignis in metallicum corpus ivisse; usque adeo ut alicubi ambientis quasi catini fusorii figuram asumsisse[27] videatur. Certe in Alabastrite Nordhusano non ita pridem argentum granulatum repertum est. Memini quasdam massulas metallicas recentes ex fodina ad me delatas, quas fusione nuper paratas esse Iovem lapidem artifices juravissent. Ita natura pro homine imponit. Contra subtiles quidam mangones mineralium rariores formas, velut argentum rude rubrum, et vitriforme et capillare, ut curiosos fallant, in foculo imitantur: ita prosunt decipiendo, docentque artem naturae, cujus effecta expressere.

[XIV]

Interea fatendum est, quaedam formas accipere solo motu aquarum, ut calorem advocare necesse non sit; quaedam indigere opera utriusque. Nil dico de silicibus torrentium diuturna provolutione tornatis, certe nihil prohibet, grana metalli cursu attrituque rotundata esse. Ut saccarata Virdunensia cum aliquandiu agitantur. Cernuntur passim politi silices in montium parietibus, ipsisque Alpibus a natura caementati, magno indicio, postquam diuturna aquarum provolutio attriverat, saxificabili terrae haesisse, post novis ruinis iterum detecta apparuisse. Unde etiam intelligitur, illic olim vel superiore alio loco flumen vel torrentem fuisse, mutata terrarum facie interversum.

26. B: accidit.
27. B: assumpsisse.

[XIII. *It is through fire that metals appear in their proper forms*]

One must assume of silver, gold, or other metals that appear unchangeable in their own form, or that are at least similar to fine gold, that they did not become metallic bodies without the force of fire. Thus, these metals seem to have taken on the shape of their surroundings, as if in a crucible. Certainly, granular silver was discovered in Nordhausen's alabaster not too long ago. I recall that someone brought me small metallic pieces, freshly mined; the adepts would have sworn to Jupiter that they had been prepared through smelting. Nature, instead of humanity, deceives. On the other side, skilled connivers imitate rare mineral forms, like coarse red, vitriform, or fibrous silver, in order to deceive the curious. They are thus useful in their deception, and teach the art of nature, whose effects they copy.

[XIV. *Some bodies owe their form to the movement of waters*]

Nevertheless, one must admit that some things assume their form solely through the movement of the waters, so that it is not necessary to include the heat, while others demand the action of both. I do not want to speak of pebbles, which have been rounded through prolonged rolling in the streams. And certainly, nothing keeps grains of metal from being rounded through movement and rubbing, like the sweets of Verdun when one shakes them for a while. Sometimes one sees flattened pebbles in mountain walls, and one finds them naturally cemented in the Alps, a clear proof that petrifying earth stuck them together after long rolling in the waters had abraded them, and that they became visible again after being uncovered by new collapses. One also recognizes from this that once, in that spot or a higher place, there was a stream or a torrent that moved elsewhere with the transformation of the earth's surface.

[XV]

Quin et concrescentia quaedam in mediis fluctibus non eo minus regularem figuram assumunt. Ita memini videri velut orbem quendam exiguum ex terra fictilem inter rupes Sueciae ante Stockholmam, quas *Scaras* vocant, in mari natum. Nam ad festucam aut paleam aut simile quiddam adhaerescens materia ab aqua advecta natansque, insensibiliter incrementa assumit in circuitum; adeo non disturbante fluctu, ut potius motu suo tornasse censeatur, dum aqua fortiori gyratione a centro recedens, innatantem materiam in medium contrusit.

[XVI]

Guttis cadentibus in cavernis tophaceum lapidem relinqui constat. Et in nostrae Hercyniae celebri antro, quod a Baumanno quodam nomen accepit, ingentes illas columnarum massas, percussu sonitum campanae aemulantes, et totam cavernae materiam vastosque[28] parietes quasi guttatim collectos[29] *Tropfstein* vocant,[30] id est Stalactitem, quod Spathi[31] mollioris genus videtur, stiriarum specie; quanquam in ipsa caverna non deorsum tantum respiciant, sed et in latus saepe, immo et in superiora acuminentur; nec semper, velut ex altiore loco, accumulatae consistant partes, sed (ut ruptis fragminibus apparuit) radios, folia, strias in omnes plagas protendant. Unum ipse frustum de caverna in lucem proferri mecum jussi, cumque acuratius[32] inspicerem, ecce ostendunt sese alveoli quidam in ipsa saxi superficie ad cavernulae modum introrsum recedentes, quales in Spatho fodinarum detectos et saepe peramplos metallicolae nostri *Drusas* vocant. Vestiuntur intus ambientibus totam parietum superficiem ingemmamentis: sic enim merito appellaveris illas acumina sibi obvertentes pseudocrystallos adamantiformes. Itaque non satis scio, an dicere

28. B: vastasque.
29. B: collectos, huc referre licet.
30. B: Vulgo *Tropfstein* vocant.
31. B: et Spathi.
32. B: accuratius.

[xv. *Some bodies coalesce in the waters*]

Some things actually grow together in the midst of waves and assume a regular form. For example, I remember how, between the Swedish cliffs before Stockholm, called *Scaras*, one sees something like a small circle, made of earth, that was formed in the sea. For the floating clingy material, which is carried in by the water, and is like straw or chaff or something similar, grows gradually and imperceptibly all around. The tide does not disturb this at all. On the contrary, it seems to have fashioned this rounded form through its motion, since the water, receding in ever stronger eddies from the center, has forced the floating material into the middle.

[xvi. *Kinds of tuff stone formed by dripping water*]

It is well known that falling drops in hollows leave behind tuff. The enormous columnar masses in a famous Harz cave (named after a certain Baumann), which sound like bells when they are struck, and the entire material of this cave and the vast walls, which likewise have been formed drop by drop, are called *Tropfstein*, or stalactite. It appears to be a kind of soft spar that looks like icicles. The stalactites in this cave are, however, not all pointed downward; rather, they often taper to the side and even upward. Moreover, the parts are not always formed from above. Instead (as one notices from the broken pieces) rays, leaves, and stripes extend in all directions. I had one piece carried out of the cave and into the light. When I examined it more carefully, cavities appeared on the surface of the stone, which withdrew into the interior like little caves, similar to what one finds, often on a very large scale, in spar from the mines, where our miners call them *Drusen*. They are gem-encrusted across the whole surface of the inner wall, for this is the name that the diamond-shaped pseudocrystals, with their tips pointed at one another, merit. I do not rightly know, therefore, whether one can claim that these caves, covered on all sides, were formed by slowly falling drops of water; for such would be more likely to result in a continuous pile, formed grain by grain, not even to mention that the stone left behind by flowing or dripping water tends to be much less hard than ordinary rock, and more like tuff. Thus, in the town of Langenholtensen, near Alfeld, there is a stream whose water

liceat, has cavitates quaquaversum incrustatas a labentibus diu guttis factas; ex quibus facilius orietur aliquid, quod velut arenulatim concrescens unum continuum cumulum componat. Ut taceam, longe mollius esse saxum solere, et topho simile, quod aqua fluens stillansque reliquit. Ita prope Alfeldam, in villa Langerholtensen, rivus est, qui aqua materiam crassiorem secum ducit, quae in lapidem cinereum topho similem versa rotas molendini aliaque, quibus allabitur, paulatim obducit. Et similem crustam candidam alicubi Salificia stramini circumdant, per quod muria tenuior decurrere jubetur, ut a sole et aere vice coctionis aquae pars consumitur; nostri leccaria opera *Leckwerk* vocant, ignotam antiquis artem. Credibile est quam causam cavernula ingemmata habuit in exiguo saxo, eam subinde ingentia antra in ipsis montibus produxisse. Sane mediocria passim occurrunt. Et contingit interdum, ut terebrae fossorum metallis insistentium in caecum hujusmodi foramen saxo conclusum delatae operam ludant, neque enim pyrii loculi tunc recte adhiberentur, quibus rumpi saxa solent; neque in laxiore spatio ictus ratio sibi constaret. Aliquando aqua intus inventa est, imo et aperto aditu noxii non raro halitus prodierunt. Et scimus ante annos non ita multos vaporem erupisse, qui flamma ad lucernam concepta operarios foede ambussit. Illud quoque dignum memoratu bufones aliquando reperiri in profundissimis puteis ipsoque saxo, vivos quidem, sed adeo stupentes, ut multorum mensium spatio vix locum mutasse deprehendantur.

[XVII]

De Baumanni antro, quod miliaribus aliquot abest a fodinis, mox rursus erit dicendi locus. Nunc illud absolvamus, ut[33] quaedam igni soli, alia soli motui aquarum et sedimentis debentur,[34] ita interdum caloris et aquae junctas operas requiri; alicubi variantibus causis ambiguum judicium esse.[35] Nam scimus quaedam ex solvente praecipitari non pulveris, sed metalli jam sui forma, et quasi fusionis imitamento subsidere: adeo ancipitis consilii est de rerum generationibus ex vultu externo pronuntiare.

33. B: ut quemadmodum quaedam.
34. B: deberi diximus.
35. B: esse declaremus.

carries a denser material, which forms a gray stone like tuff that gradually covers the mill wheels and other things that it bathes. In some places,
salted waters coat with a similar white crust the straw, through which
one lets a thinner salt solution drain, so that a portion of the water will
be consumed through sun and air instead of through boiling. We call this
operation of graduation *Leckwerk*, an art unknown to the ancients.[39] In all
likelihood, the same cause that produced shimmering cavities in small
precious stones generated large caves in the mountains. Actually, some
of a middling size exist as well. And it has happened that the gimlets of
the miners, searching for ore, have stumbled upon such a hole hidden in
the rock, so that all work is hopeless, because neither can they properly
apply the powder with which one generally blasts the stones, nor would
the action of the explosion be effective in a larger space. Sometimes water has been discovered inside, and harmful vapors have not infrequently
come out of the opening. Thus, we know that, not too many years ago, a
gas that was ignited by the flames of the lanterns erupted and burned the
workers horribly. It is also worth noting that one sometimes finds frogs
in the deepest shafts and even in the rock; these are, indeed, alive but so
torpid that they seem hardly to have moved in many months.

[XVII. *Some things arise from the combined
action of heat and water*]

There will soon be an opportunity to speak again about the Baumann Cave,
which is some miles from the mines.[40] Now we want to establish that, as
certain things owe their formation to fire and others purely to the movement and deposits of the waters, sometimes the combined action of heat
and water is required; where the causes vary, the verdict is ambiguous.
For we know that some things are precipitated by solvent not as a powder,
but in the form of a metal, and settle as if in imitation of melting, so that
it is dangerous to pronounce about the generation of things on the basis

39. Saline works.
40. Cf. *Protogaea*, §XXXVII.

Attente omnia excutienda sunt, ut constet, quid aquis tantum, quid calori tantum, quid utrique tribuendum; utrum sicco opere fusio aut sublimatio rem peregerint; an humidi solventis interventu, praecipitatio aut crystallismus dominentur: quando crystallos sublimationibus, fusionem praecipitando suppleri dictum est. Et velim microscopia ad inquisitionem adhiberi, quibus tantum praestitit sagax Leewenhoekii, Philosophi Delphensis, diligentia, ut saepe indigner humanae ignaviae, quae aperire oculos, et in paratam scientiae possessionem ingredi non dignatur. Nam si saperemus, jam ille[36] imitatores haberet.

[XVIII]

Sed illustriora omnia exemplo erunt inquisitionis in memorabile apud nos opus naturae, super scissili lapide aerosas piscium formas exhibentis. Nimirum in Islebia, Saxoniae oppido, ditionis Mansfeldicae, et nuper prope nostram Hercyniae Osterodam,[37] effoditur lapis niger foliatus, quem merito, (licet alio quam vulgo sensu) schistum appelles; quidam *Ardosiam*,[38] semilatino verbo vocant. In eo crebrae visuntur piscium formae, exacte et a fabre[39] delineatae, quasi artifex nigro lapidi sectilem metalli materiam inseruisset, ichthyomorphum aliqui indigitant. Constat Ardosium[40] fissilem esse et ex foliis ac velut tabulis constare. Teutonice vocamus Lagen, unde et lapidi huic nomen Layae est, apud Superiores Germanos; quod et illustri Familiae de Petra vernaculum, quae rarissimo exemplo, nostro tempore duos ad Rhenum Electores fratres simul dedit. Talis lapidis unum mihi fragmentum est ab utraque parte piscis imagine, sed diversa, signatum.[41] Reperiuntur in pendente vena, nam ubi terra suprema argillacea et proximae ab ea rupes perfossae, varia schisti aerosi strata Islebiae occurrunt, sed in uno tandem pisees; idque

36. B: jam passim ille.
37. B: Hercyniae municipium.
38. B: *Ardesiam.*
39. B: affabre.
40. B: ardesiam.
41. In Hannover Ms XXIII, 23b, pl. 2 appears here.

of their outward appearances. One ought to examine everything carefully, so that it is certain how much to assign only to waters, how much to fire alone, and how much to both together, whether the thing has traveled to dryness through melting or sublimation, or whether a wet solvent intervened and settling, or crystallization predominated. For it was said that sublimation can replace crystallization and that deposition can replace fusion. I also wish that the microscope, with which the Delft philosopher Leeuwenhoek has shown so much wisdom and care, were used for this investigation.[41] And I am often upset by the idleness of humans, who do not bother to open their eyes and take possession of an already completed science. For if we were that clever, he would already have many imitators.

[XVIII. *Where do the shapes of various fish imprinted on slates come from?*]

But everything will be clarified by a local example: the investigation of a remarkable work of nature that produces the coppery shapes of fish upon schistous stone.[42] Namely, in Eisleben, a Saxon town in the region of Mansfeld, near our Harz town of Osterode, a black foliated stone is mined, which is properly called slate (though in another sense than the common one), and which some, using a half-Latin word, call *Ardosia*. One often sees in these stones, which some call ichthyomorphic, the shapes of fish whose contours have been traced precisely, as if an artisan had inserted carved metallic material into the black stone. It is well known that *Ardosia*, which is easy to split, consists of sheets and also of tablets. In German we speak of *Lagen*, which is why this stone is called *Laya* in upper Germany; this is also the common name of the famous de Petra family, which (a very rare example) in our time simultaneously produced two brothers as electors along the Rhine. I own a fragment

41. Antoni van Leeuwenhoek had discovered "spermatic animalcules" with the help of a microscope in 1675. Cf. Leeuwenhoek 1677–1678.

42. Here begins a long discussion, mostly inspired by Steno, on the nature and meaning of "fossil objects." The belief that they were "figured stones" or "games of nature" was widespread during the sixteenth century, but by 1692 the organic nature of "fossil objects" was accepted by many European scholars. Leibniz finds it necessary to take up the whole argument again, probably because his own "conversion" to this "truth" was relatively recent at the time of writing. Cf. Rudwick 1972; and Cohen 1998.

genus prae caeteris ignem meretur, neque ulla aeris minera facilius fusori paret. Crassities strati ad sedecim pollices, interdum tamen contrahitur in tenuem laminam velut cultelli: sed tanto ditior in angustiis massa est. Durissimi lapidis parietes utrinque claudunt venam. Habui ipse ibi in manibus mugilem, percam, alburnam, petrae insculptos. Paulo ante erutus erat ingens lucius, flexo corpore, ore aperto, quasi sic deprehensus vivus gorgonia vi obriguisset. Visi et marini generis, ut raja, halex et lampreta, et haec aliquando halece decussata. Hic plerique ad lusos naturae confugiunt; qui et ichthyomorphi, nostri lapidis exemplo usi, tanquam indubitato oblectantis se rerum genii specimine, etiam caetera controversa firmare sperant, in quibus magnam archetectatricem velut per jocum, dentes atque ossa animalium, conchylia, serpentesque imitatam volunt. Quoties enim eo argumento premuntur, quod rudes tantum adumbrationes rei animalis extra animal nasci consentaneum sit, ad nostros lapides provocant, ubi fatendum est, delineationi nihil perfectionis adjici posse. Nam plerumque genus piscis agnoscas primo obtutu, neque unquam a symmetria abit animal, aut magnitudinem non habet suam. Sed vereor, ne, uti nimis validi ictus in autorem repercutiuntur, ita nimiae similitudinis argumentum in contrarium valeat. Tanta piscium simulatorum cum veris convenientia est, pinnis ipsis squamisque ad minutias usque expressis; tantaque imaginum frequentia in eodem loco visitur, ut manifestiorem constantioremque causam suspectemus, quam aut casum ludentem aut seminales, nescio quas ideas, inania philosophorum vocabula. Quid ergo si lacum ingentem cum piscibus suis, vel terrae motu, vel aquarum vi, vel alia magna causa dicamus obrutum terris, quae deinde in lapidem duratae impressi piscis vestigia, et velut ectypos, molli ante massae impressos, et postremo consumtis dudum animalis reliquiis metallica materia oppletos servarunt? Nec jam valde disputo, quae causa ex terra ardosium[42] lapidem fecerit, adduxeritque metallum in cavitates. Fieri potest, ut quemadmodum in fornacibus ex argilla coctiles lapides humana arte formantur, ita magnus naturae ignis pro variis terrarum generibus atque misturis nunc ardosium,[43] nunc alabastritem fecerit, aliudve petrae genus; metallica interim materia, quae toto limo dispersa erat, vi caloris exudante,[44] et in cavitates illas sese colligente, quas reliquerat piscis. Hujus enim volatilis materia dudum tempore aut calore consumta facile

---

42. B: ardesiam.
43. B: ardesiam.
44. B: exsudante.

of such a stone, each side of which is imprinted with the image of a differ-ent fish [fig. 3.]. They are found in a hanging vein, since after excavation of the superficial clay earth and the subsequent rocks, there occur in Eisleben various layers of coppery schist. But only one layer has fish, and this one is especially well suited to the fire, for no other copper ore obeys the smelter more easily. This layer is about sixteen inches thick, though sometimes it gets as thin as a knife blade; but the narrower the mass, the richer it gets. The vein is enclosed on each side by walls of the hardest stone. I have here in my hands a barbel, a perch, a bleak, sculpted in stone. Not long ago an immense pike was dug out of a quarry, its body bent and its mouth open, as if it had been caught alive and turned to stone by the power of the Gor-gon. I have also seen sea fish like the ray, the herring, and the lamprey, the last one sometimes lying crosswise with a herring. Here most take refuge in games of nature, trying to use our ichthyomorphic stones as an indu-bitable example of the playful genius of nature, and hope thereby to re-solve other controversies, in which they claim the great architect, as if in jest, had imitated the teeth and bones of animals, shells, or snakes. When pressed with the argument that, reasonably speaking, only crude outlines of an animal could arise outside the animal realm, they adduce our stones as evidence, where one has to admit that nothing can be added to the per-fection of their shapes. In most cases, the kind of fish can be recognized at first glance, and the animal lacks neither symmetry nor proportion. But I am afraid these overly strong blows fall back upon their authors, since the argument of close resemblance supports just the opposite posi-tion. For the imitated fish perfectly resemble real fish, right down to the finest details of their fins and scales. Such a great number of these images is seen in the same place that we suspect a more obvious and uniform cause than playful accident or some ideas about generation, the empty words of the philosophers.[43] But what if we suggest that an immense lake, together with its fish, was filled with earth, either by an earthquake, or by the force of the water, or by some other great cause; that when the

43. That is, the arguments of Athanasius Kircher, Johann Joachim Becher, and their followers. Cf. Ariew 1998, 283; and Cohen 2002, 53–54. The belief in the "generation of stones" was widespread during the Middle Ages and the Renaissance. The idea that "seeds" could have produced petrified organisms, or fragments of organisms, was ar-ticulated in the seventeenth century by scholars such as Michael Ettmüler, Alphonso Borelli, and Ferrante Imperato, and at the beginning of the eighteenth by Herman Boerhaave and Joseph Pitton de Tournefort, among others. See Adams 1938.

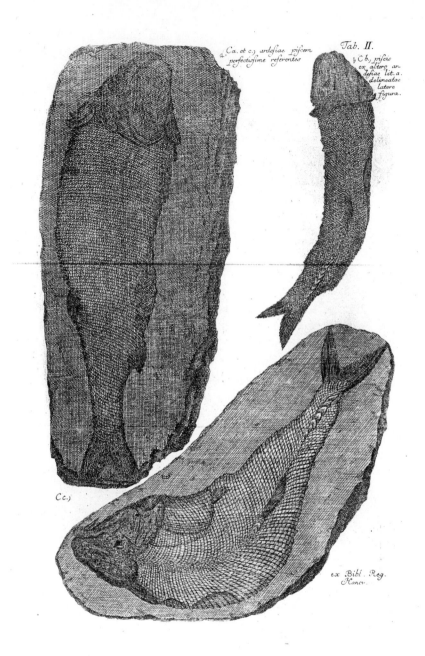

FIGURE 3. *Fish Imprinted on Slate*

This plate shows several "ichthyomorphic stones," that is, images of fish imprinted on slates. According to Leibniz, the similarity between these stones and real fish offered the best proof of their organic origin, enabling him to reject the notion that they resulted from "games of nature" (see *Protogaea*, §XX).

This unsigned engraving reproduced specimens from the Royal Library in Hannover. The original captions marked "Ca" and "Cc" read, "Slates bearing the most perfect fish"; the caption marked "Cb" reads, "Side view of a fish from another slate." This figure is reproduced from Scheidt's 1749 edition (as are the figures from 2 to 12), where it appeared as plate 2.

locum fecit. Habemus aliquid simile in aurifabrorum artificiis; libenter enim occulta naturae manifestis hominum operibus confero. Araneum aut aliud animal materia apta obruunt, aditu tamen exiguo relicto; materiam igne coquunt in lapidem, inde hydrargyro immisso cineres animalis per foramen eliciunt, postremo pro illis eadem via argentum infundunt; ita demum remoto putamine inveniunt argenteum animal omni pedum et capillamentorum apparatu, fibrillisque ad stuporem assimilatis.

[XIX]

Neque adeo mirari oportet calorem terras coquentem in lapidem, aut metalla in minerales massas fundentem, aut materiam in figurata corpora sublimantem, aut remittente solutione atque aestu deponentem in crystallos; cum plerique credant,[45] ignem esse inclusum huic globo, cujus vix crusta nobis explorata est; et terrae motus[46] validissimi pyrios intus cuniculos innuant, et Vulcani igentes[47] late patentia pyrophylacia ostendant. Certe nupera terrae concussio anni 1691[48] ab Italia ad nostros usque fines pervenit, quanquam Visurgim non transgressa. Habentur et passim fodinae lithanthracum, materiaque sulphuris, quae et mineras nostras

45. B: cum non solum eruditorum plerique.
46. B: Sed terrae etiam motus.
47. B: ingentes.
48. B: anni hujus 1691.

earth later hardened to stone it was imprinted with the remains of the fish which, like copies, had initially been stamped upon the soft mass; and finally that, after the remains of the animals were long gone, the spaces they left were filled with metallic matter? I will not discuss here in detail what turned earth into *Ardosia* and drew metal into the cavities. It could be that nature's great fire formed schist, alabaster, or other kinds of rock according to the different kinds of earth and their mixtures, just as bricks are formed out of clay in the ovens through human art. Meanwhile, the metallic matter, which was scattered all through the mud, was separated out by the force of the heat, collecting itself in the cavities left behind by the fish. For the volatile material, long since consumed by time or by heat, easily provided room for it. We find something similar in the art of the goldsmith, for I gladly compare the secrets of nature with the visible works of men. They cover a spider or some other animal with suitable material, though leaving a small opening, and bake this material to stone in the fire. Then, by pouring in mercury, they drive the animal's ashes out through the hole, and, finally, they pour silver in the same way. When the shell is removed, they uncover a silver animal, with its entire complement of feet, hairs, and fibers, which are wonderfully imitated.

[XIX. *Earthquakes, volcanoes, and other things show that there is fire inside our globe*]

Nor should one wonder that heat turns earths to stone, that it melts metals into mineral masses, that it sublimes matter into fashioned bodies or deposits it as crystals when the heat of a solution is reduced. For most believe that there is fire contained in this globe, whose crust we have hardly explored. Earthquakes also may clearly indicate that there are tunnels of fire, and huge volcanoes reveal fire dungeons extending far and wide. The recent earthquake of 1691 reached from Italy to our borders, though it did not cross the Weser River. One finds, here and there, deposits of pit coal and sulfuric material, which also fill our Goslar ores. There is also beautiful native sulfur—transparent and the color of ruby or gold—that is extracted somewhere in our region of Lauenstein, as if it had already encountered the fire of nature, even though we call it *Apyron*.[44] Nor is it

44. From the Greek ἄπυρος, "without fire."

Goslarienses implet. Et alicubi nativum sulphur pulchellum et translucidum rubini aut auri colore ex nostro Lauensteinii tractu eruitur, quasi jam ignem naturae expertum, etsi nobis apyron dicatur. Nec absurdum est, privata incendia etiam magno diluvio posteriora extitisse, ignota annalibus, cum materia combustilis adhuc per terram uberius disseminata esset. Pumices esse ex locis qui arserunt, Agricola merito judicat, nec in Sicilia tantum et Campania, sed etiam in Germania constat reperiri. Ipse Agricola apud Mosellae confluentes et Grani Aquas talia agnoscit. Naphthae autem aut bituminis liquidi fontes etiam apud nos fluunt; quale est prope Burgdorfium, quod rustici in axungiam absumunt. Arsisse hos tractus Agricola (credo post Cordum nostratem) non dubitavit. Nam cum Belemniten, quem aliqui Lyncurium vocant, in ripa Leinae amnis ad Neapolin oppidum duodecim ab Hannovera passum millibus, et in Brunsvicensi tractu, ubi bitumen gignitur, et potissimum inter urbem Hildesiam, et arcem Marienburgum in marmore antri, quod a nanis appellant, reperiri dixisset, hunc locum, inquit, arsisse multa indicant. Idem mox repetit, cum Ostraciten, ita dictum a testae similitudine, circa idem Hildesheimii antrum reperiri narrasset. Cui consentit, quod superius annotaverat: Natura, inquit, ex Ostracite in Hildesheimio tractu facit haematiten, ars ex magnete; utraque urendo.

[XX]

Si quis tamen coctile naturae saxum aegrius admissurus malit, limum pisces obvolventem vel ipso tempore ex natura materiae, vel aliunde, lapidifico quodam spiritu, aliave causa, in petram abiisse, et materiam metallicam in piscis modulos, aut initio cruda et molli adhuc massa, aut postea etiam penetrabili vapore advectam (quanquam haec minus facile intelligantur) non repugno, nec aliquid certe constituere audeo, nisi quod nobis satis est, pisces aerosos ex veris expressos. Firmant sententiam et multidudo piscium uno eodemque loco conclusa, et quod ibi nil nisi pisces. Nam quae de Triregno Pontificio, de Luthero, de nescio quibus aliis formis in petra Islebiensi delineatis jactant, haec vere inter lusus habeo, non jam naturae sed imaginationis humanae, quae in nubibus acies videt, et in campanarum aut tympanorum pulsibus quas vult modulationes

unreasonable to suppose that particular fires, unrecorded in our histories, occurred after the Great Flood, when combustible material was more abundantly spread across the earth. Agricola was right that pumice stones come from places that have burnt, and it is well known that they are found not only in Sicily and Campania, but also in Germany. Agricola himself identified them at the mouth of the Mosel and along the Hron River.[45] Even springs of oil or liquid bitumen also flow in our region. This is the case near Burgdorf, where peasants use it to grease their wheels. Agricola (following our own Cordus, I believe) did not doubt that these regions had burned.[46] For after declaring that belemnites, which some people call lynx stones, were found along the banks of the Leine River near Neustadt, twelve thousand double steps[47] from Hannover in the region of Brunswick (where bitumen is produced)—and above all that they were found between the city of Hildesheim and the Castle Marienburg in the marble of the so-called Cave of Dwarves—he then says that there is much to indicate that this region once burned. He repeats the same thing not much later, after having reported that one finds ostracites (so called because of their resemblance to oyster shells) in the vicinity of the same Hildesheim cave.[48] This agrees with what he noted earlier: nature, he says, makes a hematite from an ostracite in the region of Hildesheim, just as art makes hematite[49] from a magnet; in both cases, this happens through burning.

[xx. *The forms of fish imprinted on slate come from real fish, and are not games of nature*]

If, however. someone does not want to accept that nature burns rocks, and prefers to think that the mud enveloping the fish turned to stone, either through time alone and according to the nature of the material,

45. Cf. Agricola [1546] 1955, bk. 5.

46. Valerius Cordus (1515–1544), German physician and botanist. See Ogilvie 2006, 34–35, 145–147.

47. One *Doppelschritt* (double step) was about 148 centimeters, which makes this about eleven miles. In fact, traveling from Neustadt to Hannover along the Leine covers a distance of about thirty-three kilometers, or twenty miles.

48. Cf. Agricola [1546] 1955, bk. 5.

49. Hematite, or "bloodstone," is a reddish mineral composed of iron compounds. See glossary.

agnoscit. Et talia sunt multa, quae vulgo in Baumanni antro ostendunt: Mosem scilicet, et Ascensionem Christi, aliasque ex lapide figuras, quas nisi admoneare, non agnoscas. Augere rerum species in miraculi fidem, ut stupenda de nostris regionibus dixisse videar, non est meum. Sed piscium Osterodanorum atque Islebiensium summa est manifestaque in repraesentando fides; nec piscem tantum sed et genus piscis et veram magnitudinem et commensurationes partium, et squamas, et caetera omnia statim fatebere. Accedit magnum argumentum ab ipsa constitutione loci. Nam venam ardosii pisciferi pendentem esse diximus, ut cum nostris fossoribus loquar; id est in strati prope horizontalis modum procurrentem per aliquot miliaria, ut jam facile appareat, ejusdem stagni pisces a superincumbente mole pressos. Certe in vicino Islebiae agro insignes lacus nunc quoque extant. Et quo minus marinos in petra pisces mirere, non procul abest Seeburgum, ubi ingens aquae salsae stagnum; et sub terra esse conditiora salis fontes aquarum salsarum docent; ex quibus illustris inprimis est, quem Hala Saxonum octava ab Islebia lapide fundit, quem Chattis olim et Hermunduris materiam bellandi quidam putarunt.

or through some petrifying spirit, or through another cause, and if one wants to assume that the metallic material was driven into the molds of the fishes, either in the beginning when the mass was raw and soft, or also later as a penetrating vapor, then I do not oppose it, though I find it harder to understand. I do not dare to assert anything with certainty, except one thing, which suffices for us here: namely, that coppery fish are the imprints of real ones. This opinion is supported by the fact that there are many fish enclosed in the same place, and that there are nothing but fish there. As to the supposed appearance of the pope's tiara, of Luther, and all sorts of other shapes etched in the stone of Eisleben, I consider these to be games not of nature, but of the human imagination, which sees battles in the clouds and hears its favorite melodies in the sound of bells or the beating of drums. And of this kind is much of what is commonly displayed in the Baumann Cave, like Moses, the Ascension of Christ, and other stone figures that you would not recognize unless you were forewarned. It is not my purpose to increase the varieties of such things, or to make miracles out of them so as to say extraordinary things about our regions. But the best and most obvious proof comes from the depiction of fish in Osterode and Eisleben; one recognizes immediately not only the fish, but also the kind of fish, its true size and the dimensions of its parts, its scales, and all the rest. In addition, the disposition of the place provides an important argument. For we said that this fish-bearing schist vein is a hanging vein, to speak as our miners do;[50] that is, it extends for several miles like a horizontal layer, from which it would clearly appear that the fish in one and the same lake were buried by a covering mass. In fact, there are still prominent lakes near Eisleben. Nor should one wonder at the existence of petrified saltwater fish, since there is an immense sea with salt water not far from Seeburg;[51] moreover, the saltwater springs demonstrate that there is salt buried beneath the ground. Especially well known among these springs is the one that flows at Halle in Saxony, which is eight miles from Eisleben. Some believe it was once the cause of war between the Chatti and the Hermunduri.

50. Cf. *Protogaea*, §VIII.

51. The Seeburger See, roughly eight hundred meters wide and five kilometers long and one of the largest lakes in Lower Saxony, supports a large variety of fish to this day.

Hinc iam intelligas duplicem, ut ita dicam apud nos terram; una[49] cum pisces in suo lacu essent, alteram postquam obrutis opposita est ingens materiae mollis moles, quae tandem sive igne, sive gorgonio vapore aut salibus, sive tempore ipso in diversa lapidis scissilis strata secessit; cui impositae rupes durissimae, atque his demum argilla superjecta, ac postremo communis terra nigricans nata est, quam nunc homines colunt. Et licet velimus, lacus illos antequam obruerentur subterraneos fuisse, necesse est tamen supremam terrae superficiem, quae tunc fuit, prorsus convulsam commutatamque intercidisse. Quidam Hispanus de duplici viventium terra olim conjecturas edidit. Sed omnia altius expenderat vir egregius, qui de solido intra solidum scripsit, eoque[50] inclinabat: ut olim per summorum montium iuga telluris planities incesserit fractos deinde fornices, ubi minus firmi erant, in inferiora procubuisse, inclinata passim strata initio ad libellam composita, casum testari.[51] Quin et credibile est, aliam super priores ruinas structuram stratorum horizontalium natam, ex superfuso inter ruendum liquido limo post indurato. Quae strata et ipsa deinde posterioribus ruinis interrupta apparent. Ita tres quasi telluris contignationes statuemus; summa montium juga, colles medios, tractusque imos litorales. Mare autem olim et in excelsissimis locis fuisse arbitrabatur ille, passimque ejus vestigia habentur. Et consentaneum est, ruptis fornicibus defluxisse cum ostreis limum fundi, et in mediis locis ab aqua in ima penatrante desertum haesisse, postremoque in lapidem abiisse. Unde plena conchyliis strata visuntur. Et facile fieri potuit, ut aqua salsa alicubi intercepta haereret in cavernis vel pura vel terrae fluitanti permista, unde dissipato humore vel sal gemmeus superfuit, ut in Cracoviensibus fodinis, vel terra sale gravida, ut in Tirolensium Alpibus, unde Halenses illi immissa aqua dulci eliciunt vires. Idemque naturam facere credibile est, cum muriae rivos emmittit. Nam aquam pluviam aut nivalem facere fontes, pro comperto est, qui deinde per salis gemmei rupes aut terram saturatam in montium angustiis fluentes, assumto sapore in

49. B: unam.

50. B: Eo enim inclinabat, ut olim per summorum montium juga telluris planitiem incessisse.

51. B: testari docuerit.

[XXI. *The different layers of the earth, their locations, and the origin of salts and salt waters*]

From this you now understand that around us the earth has two tiers, so to speak: the one was formed while the fishes were in their lake; the second arose after an immense mass of soft material covered them and then collected upon them. It finally hardened into several layers of splittable stone, whether through fire, petrifying vapors, salts, or time itself. These layers were covered with very hard rocks, then with clay, and finally with common black earth, which people cultivate today. Even if we suppose that these lakes were under the earth before they were filled in, we would still have to recognize that the onetime surface of the earth had been entirely transformed and ultimately scraped away. Some time ago, a certain Spaniard published conjectures about the two-tiered earth of living things.[52] But that eminent man who wrote on "solids enclosed within other solids,"[53] considered everything better. He inclined toward the conclusion that the smooth surface of the earth had once stretched across the highest mountain peaks; that the vaults then broke at their weak points, and sunk into the depths; and that evidence of this collapse is to be found everywhere in the twisting of the layers of the earth, which were initially formed at the water level. It is also likely that another series of horizontal layers, composed of the liquid mud that poured down during the collapse and hardened afterward, rose above the initial ruins. These layers were themselves clearly broken apart by later collapses. Thus, we can distinguish, so to speak, three floors of earth: the mountain ridges form the highest floor, the hills the middle, and the coastal regions the lowest. Steno also believed, however, that the sea had once reached even the highest places, and we have signs of it here and there. And it is likely that after the vault had broken, the sea-bottom mud flowed away, together with the shells, and clung to the places of middling height, where it eventually turned to stone, while the water left it for the depths. One thus finds layers full of shells. And it could easily happen that salt water, either pure or mixed with silty soil, was trapped somewhere and caught in the hollows. Later, after the wetness had dissipated, it left behind either rock salt, as in the Krakow mines, or soil impregnated with salt, as in

52. González de Salas 1650.
53. Steno 1669, 1.

lucem erumpunt. Certe, quemadmodum dictum est, Hala, quae in Saxonia est, habet vicina sibi stagna salsa, et non procul a Luneburgensibus salinis dentes monstrorum marinorum eruuntur.

[XXII]

Scio quosdam suspicari intumuisse aliquando terram ab erumpente spiritu, surrexisse montes ex planitie, erupisse insulas ex mari; qualis apud Cedrenum, et in historia miscella memoratur insula nata sub Leone Iconomacho. Ventorum certe quanta sit vis, Typhones declarant, et tubae marinae aquam haurientes navesque evertentes, et qui nuper turbo in Veneto tractu homines rapuit, plumbeaque tecta ad millia multa passuum tulit. Sed nihil illis terribilius est, quos in Americae insulis *uracanos* vocant, qui brevi tempore gyrum facientes plagarum mundi, ingentem in aere luctam testantur, cujus solito illic statoque cursu interrupto mare confundunt, terras evertunt. Sane pagum integrum terra vento advecta oppletum vidit Cardinalis Bellarminus. Rhodani aquam ventos stitisse, ut supra in cumulum suspenderetur, intra sicco alveo transiri posset, memorat historia Genevensis, ut credibile sit, lapides veros, qui e caelo cecidisse creduntur, ventis venisse. Nam si quis in aere coagulatos putet ex vaporum materia, quaero an triticum quoque in nubibus natum arbitretur, quo pluisse praeter veteres memorant Carinthii prope Victringam Cisterciensis disciplinae coenobium. Ego ut facile admittam, initio cum liquida esset massa globi terrae, luctante spiritu superficiem varie intumuisse, unde illi mox indurescenti primaeva inaequalitas; neque etiam diffitear, firmatis licet rebus, terrae motu aliquando vel ignivoma eructatione, monticulum factum. Sed ut vastissimae Alpes ex solida jam terra, eruptione surrexerint, minus consentaneum puto. Scimus tamen et in illis deprehendi reliquias maris. Cum ergo alterutrum factum oporteat, credibilius multo arbitror, defluxisse aquas spontaneo nisu, quam ingentem terrarum partem incredibili violentia tam alte ascendisse.

the Tyrolean Alps, where the people of Hall extract it by bathing the soil with fresh water. Nature probably does the same thing when it flushes salty ponds with brooks. Indeed, it is well established that rainwater or snow form springs; and when these springs flow into mountain crevices and over rock salt or salt-laden earth, they acquire a salty taste by the time they emerge into the open. In any case, as has already been said, there are salty lakes near Halle in Saxony, and one finds the teeth of sea monsters not far from the salt springs of Lüneburg.

[XXII. *The origin of mountains and hills explained through waters, winds, and earthquakes*]

I know that some suspect the earth was swollen by the bursting forth of the wind, and that mountains rose up from the plain, and islands erupted out of the sea. Such an island, which is mentioned by Cedrenus and in the *Historia miscella,* arose during the reign of Leo the Iconoclast.[54] Typhoons and sea tornadoes, which swallow up the water and capsize boats, show how great the force of the wind can be. Recently, in the region of Venice, a whirlwind snatched men away and carried lead roofs several thousand feet. But nothing is more terrible than what, on the islands of America, are called hurricanes, which in a short time spin in every direction of the compass and produce an immense wrestling match in the air. If their usual and regular course is interrupted, they lash the sea and gouge the land. Cardinal Bellarmine saw a whole village buried by soil that the wind had carried. The history of Geneva recounts how winds stopped the flow of the Rhône River, piling its waters into a heap and making it possible to walk across the dry riverbed. It is also plausible that real stones, which are believed to have fallen from the sky, were carried by the wind. For if one thinks that they coalesced in the air from the material of vapors, I ask whether he believes that wheat arose in the clouds as well, since the Carinthians report, contrary to the ancients, that it rained down from the sky at the Cistercian monastery near Victringa. I gladly admit that at the beginning, when the earth's mass was liquid, the struggle of the wind inflated its surface in various ways; soon afterward, with hardening, came its youthful unevenness. I would also not deny that sometimes,

54. Leo the Isaurian (ca. 680–741 CE), Byzantine emperor (717–741 CE).

Itaque ut caetera apud nos oceani vestigia prosequamur, dicendum est de conchyliis, quibus passim et nostra saxa implentur. Jam olim Valerius Cordus, insignis Medicus Brunvicensium et Hildesiensium, a quo plurima accepit Agricola, de nostris fossilibus, in Hannoverae et potissimum Hildesii latomiis et puteis, fossisque ipsis et cellis observavit rem frequentem esse, idemque juxta Alfeldam notari.[52] Et longe altiori loco, prope Grundam, oppidum metallicolarum, ex adverso Ibergi monte Hercyniae nostrae, unde praestantissima ferri minera eruitur, velut scopulus quidam (dictus *Hupkenstein*) sese attollit ex spathi genere (quale et Baumanni specus componit) in quo varia concreta conchylia visuntur. Sed idem passim tota Europa deprehendi constat. Et Figueroa, Legatus Hispanus ad Schachabassum Persam, Ormusio veniens, in excelsis Caramaniae montibus Ostrea et durissimo caemento insertas velut suae Galloeciae conchas miratus est, nec dubitavit vestigia maris fateri. Sed eadem jam dudum veteres dixere, neque hujus loci est compilare pervulgata. Praestat rem ipsam intueri, et manifesta sepulti animalis argumenta agnoscere.

52. B: observavit, marina in nostra regione frequentia esse, eademque juxta Alfeldam inveniri.

after things firmed up, a small mountain was formed by an earthquake or by a volcanic eruption. But I find it less reasonable that the mighty Alps could have risen out of the already solid earth through eruption. We know, however, that one discovers the remnants of the sea even in them. Since one or the other must have happened, it is much easier to believe that the waters sank of their own accord than that a huge part of the earth was raised so high with incredible violence.

[XXIII. *Marine shells are found throughout our region and elsewhere*]

In order to continue with the other vestiges of the ocean in our region, we have to speak about the shells that fill our stones in various places. Valerius Cordus, the distinguished physician of Brunswick and Hildesheim, from whom Agricola received most of his knowledge about our fossils in the quarries, wells, graves, and cellars of Hannover, and especially of Hildesheim, observed that this was common in our region; and he observed the same thing near Alfeld. At a much higher place, near the mining town of Grund, across from Mount Iberg in our Harz, where an excellent iron ore is excavated, there rises a cliff—the so-called Hupkenstein— made of a kind of spar (just like what holds together the Baumann Cave) in which one sees various petrified shells. But it is well known that one finds the same thing in various places throughout Europe. Figueroa,[55] the Spanish ambassador to the Shah of Persia, coming from Hormuz, was amazed to find oysters on the high mountains of Caramania[56] and, as in his own Galicia, shells encased in the hardest cement. He did not hesitate to recognize them as vestiges of the sea. But the ancients already said the same thing long ago, and this is not the place to compile things that are common knowledge. It is better to look upon the thing itself, and to recognize the obvious arguments of a buried animal.

55. Don Garcia de Silva y Figueroa served as the Spanish ambassador to Shah Abbas I, King of Persia, between 1614 and 1624.

56. To the north of Cyprus, in southern Turkey.

Primo igitur aspectu passim summa similitudo se prodit, color, lineamenta, valvae, turbines, nautili, buccinae, Echini, histricis; spinae, vertebrae, dentes animalium marinorum; et omnia ad verum modum, ut species ipsas, quas curiosorum musea ostendunt,[53] in Saxo advertas. Aliquando contusa, dimidiata, fracta et imperfecta apparent conchylia, antequam lapidi includerentur, insigni testimonio non ibi a natura, sed a casu collocata esse. Aliquando perfecta figura margae post induratae impressa testatur, integrum fuisse prototypum, quod nunc vel pro parte tantum superest, vel consumtum prorsus, ut in illis testis saxeis, quae doctis aereae aut fallaces appellantur. Testacea porro non saxo ingenita, sed limo immista, vel ipsa varietas eorum loquitur, quae in unum saepe confusa spectantur: ita ut aliquando idem frustum lapidis ex Melita et echinum, et cochleam, et milleporum, et dentem magni piscis ostendat. Visus est astacus in saxo, conchylium forcipe premens, et fragmina conchyliorum intra echinum cum marga deprehensa, minora tamen semper foramine echini, ne dubites extrorsum venisse. Aliquando materia inclusa conchylio ab ambiente diversa est, ut prius impletum, deinde translatum videatur. Neque ossea illa corpora radices agunt in matricem saxeam, filamentaque propagant, quasi illic nata; sed velut insulam faciunt sui juris, propriaque ac saepe polita superficie terminantur. Nec marga tantum, aut certa terrae genera hos velut fructus ferunt, sed saepe totum saxum, glarea et varii coloris lapilli, et steriles arenae grumi componunt, animalium spoliis interstincti, ut facile intelligas, in unum omnia cumulum pari fato collecta iisdem vinculis caementata cohaesisse.

53. B: Primo igitur aspectu passim summa similitudo se prodit, color, lineamenta, valvae, turbines, spinae, vertebrae, dentes manifeste imagines produnt nautili, buccini, echini marini, histricis et aliorum animalium marinorum, atque ita omnia ad verum modum accurate referunt, ut species ipsas, ceu curiosorum Musea ostendunt.

At first glance, the most striking similarity is obvious everywhere: the color, features, folds, twists, coils of the nautilus, of the trumpet, the urchin, and the hystrix; the spines, vertebrae, and teeth of sea animals; and all of it in true fashion, so that you see the species themselves, as they are shown in cabinets of curiosities, in the stone. Sometimes it appears that the shells were crushed, cut in half, or broken and imperfect before being enclosed in stone, striking evidence that they were not placed there by nature, but by chance. Sometimes the perfect shape, printed in clay that hardened afterward, demonstrates that the prototype, which now only remains partially, or has been entirely consumed, was once complete. This is the case among those shells that scholars call airy or deceptive. That these shells did not arise in the rock, but were mixed together with the mud, is evident from the confused variety of species that one often observes in the same place. Thus, the same Malta stone sometimes reveals an urchin, a snail, a millepore, and the tooth of a large fish. In one rock was seen a sea crab, squeezing a shellfish in its pincers, and one found shell pieces mixed with clay inside sea urchins, but they were always smaller than the urchin's opening, so that they could only have come in from the outside. Sometimes the material inside the shell is different from what surrounds it, almost as if it was first filled up and then carried away. And these bony bodies are not rooted in the stony matrix, nor do they extend filaments as if they were born there; rather, they form an island with its own law, and they are bounded by their own, often polished surfaces. And these shells are borne like fruit, not only by clay or other kinds of earth, but the whole stone is often filled out by gravel, small stones of various colors, or piles of barren sand, checkered with the remains of animals. You can easily see then that the same fate brought everything together in a single heap, cemented by the same bond. [See fig. 4.]

Tab. IV.

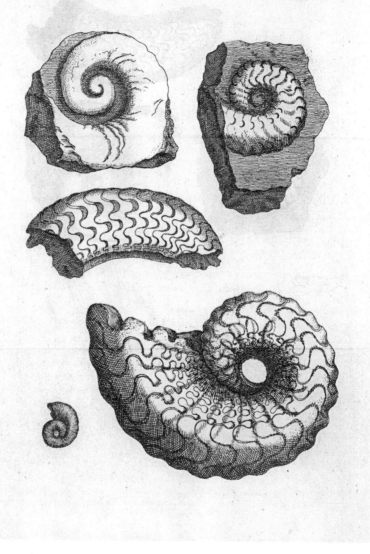

FIGURE 4. *Ammon's Horns*

This engraving depicts various "Ammon's horns," today recognized as cephalo-pods. The question of the nature, origin, and disappearance (after they were rec-ognized as animal remains) of Ammon's horns was widely discussed during the seventeenth and eighteenth centuries. Buffon took up Leibniz's argument in his *Histoire naturelle* (1779, vol. 5, 297), arguing that the animals represented by these petrified shells, no longer found in the natural world, might be living in the ocean depths. But Buffon also ventured beyond Leibniz by suggesting that they might be the remains of extinct animals.

This unsigned engraving (plate 4 in Scheidt's 1749 edition) was modeled on the sketches in Lachmund's *Oryktographia* (Lachmund 1669, 34–35).

XXV

Quanto exactius introspicies ipsas corporum partes, eo minus de origine dubitabis: neque enim refugiunt examen, ut ludicra illa in marmoribus imitamenta hominum aut domorum, quae ex longinquo spectare oportet, ut similia credantur. Analysis diligentior ostendet, non rupea minus, quam littoralia testacea eodem texturae genere ex quibusdam crustis et filamentis, et velut suturis consistere, et in cellulas distingui, in aceto quoque dissolvi, (quoties scilicet a lapidea materia magis tecta, quam penetrata sunt,) et aliqando⁵⁴ margaritas intus repertas, animalque ipsum in sua concha saxi succo, quasi balsamo conditum. Postremo prope Volaterram Tusciae, et prope Rhegium Calabriae manifestae cochleae nihil omnino mutationis praeferentes repertae sunt in terrae stratis, sine ulla petrificatione. Quemadmodum et animalium spolia apud nos ex limo eruuntur in antro prope Scharzfeldam, quod apud incolas a Pygmaeis denominatur. Nulla ergo ratio est, cur non eandem originem judicemus, ubi terra in lapidem versa est.

XXVI

Quae contra ingerunt viri docti, parum gravant. Aegre sibi persuadent quidam, mare in summis montibus fuisse, aut illic res marinas extitisse: Scilicet quia nimis ex praesenti facie aestimant veterem terrae vultum, et diluvium omne non nisi a pluviis petunt, non satis considerantes, aliquando magnae abyssi ruptos fontes exundasse. Alii mirantur in saxis passim species videri, quas vel in orbe cognito, vel saltem in vicinis locis frustra quaeras. Ita cornua Hammonis,⁵⁵ quae ex Nautilorum numero habeantur, passim et forma et magnitudine (nam et pedali diametro aliquando reperiuntur) ab omnibus illis naturis discrepare dicunt, quas praebet mare. Sed quis absconditos ejus recessus aut subterraneas

54. B: aliquando.
55. B: Cornua Ammonis.

[XXV. *The excavated shells and bones of marine animals can be identified as the parts of real animals*]

The more closely you observe these body parts, the less you doubt their origin. In fact, they bear up under scrutiny, unlike those playful imitations of people or houses in marble, whose likeness depends on seeing them from afar. A more careful examination will show that petrified shells, no less than those from the seashore, are composed of the same textures, crusts, fibers, and, so to speak, seams; that they are divided into chambers; that they can also be dissolved in vinegar, insofar as they are covered with stony material more than penetrated by it; and that one sometimes finds pearls in them, and even the animal itself, as if it had been embalmed in its shell by the juice of the stone. Moreover one has found what are clearly shells in layers of the earth near Volterra in Tuscany and near Reggio in Calabria, and they display absolutely no change and were not at all petrified. Similarly, in our region, remains of animals are dug out of the mud in a cave near Sharzfeld that the locals call the Cave of Dwarves. There is thus no reason we should not assume the same origin where earth has turned to stone.

[XXVI. *In ancient times, nearby seas contained animals and shellfish that are no longer found there*]

What the learned oppose to these observations carries little weight. Some find it hard to persuade themselves that the sea was on the highest mountains, or that marine things were there. Obviously, they judge the earth's ancient aspect too much by its present appearance, and they would explain every flood only by the rains, without considering enough that the springs of the great abyss once burst open and overflowed. Others wonder at the species one sees everywhere in stones, for which you would seek vainly in the known world, or at least in our local places. Thus, they say that Ammon's horns,[57] which many consider a kind of nautilus, sometimes differ in form and size (for some have been found that are a foot

---

57. The goatlike horns of Jupiter Ammon call to mind the spiral shape of these fossil cephalopods. See glossary.

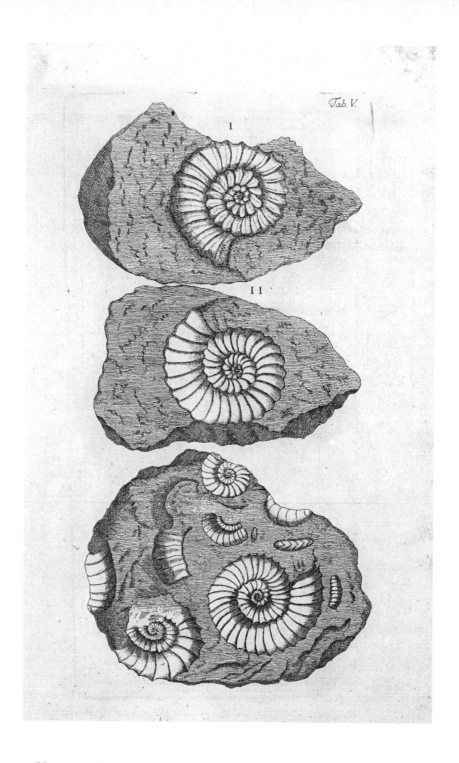

Tab. V.

I

II

FIGURE 5. *Petrifications*

The first two images illustrate a shell and its stony matrix, which fits it like a mold. This fossil object was discovered in a gravel quarry. The third image depicts a "stony mass with molds of grooved snails, partly hidden and partly visible" (Lachmund 1669). The engraving illustrated the process of petrification itself, as explained by Leibniz and Steno. The shapes of the shells, imprinted in clay that hardened later, were, as Leibniz put it, not themselves rooted in the stony matrix, but "form an island with its own law, and . . . are bound by their own, often polished surface" (*Protogaea*, § XXIV).

This unsigned engraving (plate 5 in Scheidt's 1749 edition) was modeled on Lachmund's sketches (Lachmund 1669, 49–51).

abyssos pervestigavit? Quam multa nobis animalia antea ignota offert novus orbis? Et credibile est per magnas illas conversiones etiam animalium species plurimum immutatas. Cornua Ammonis nostra exhibet Lachmundus in fossilibus patriis, unde figuras huc transtulimus.[56] Sed de his amplius disquirit diligens naturae operum investigator Joh. Raius, Anglus. Nec dubito in tanta rerum perturbatione ex longinquis oris saepe advecta maris spolia, cum nunc quoque constet, passim tempestates in littoribus ejicere genera conchyliorum, quae piscatores ex vicino mari non educunt. Et quod immensa similium massa in uno loco concrevit, quemadmodum in una Melita sub glossopetrarum nomine infinitos marinorum canum dentes miramur, non inepte aquarum vorticibus tribuas, ubi post multas agitationes ea potissimum collecta in unum deponi oportuit, quae motu ac pondere conspirabant. Prorsus quemadmodum in crystallificio salium diversorum in eodem liquido solutorum cognatis sine confusione adjungi videmus. Plerumque autem, ut arbitror, aqua per angusta loca viam reperiens, quae advehebat, destituit. Ita necesse fuit, novo semper oneratoque spoliis affluxu mox cis angustias ingentem marinarum rerum molem accumulari.

[XXVII]

Postremo magis magisque in dies observationibus naturae consultorum vera lapidum prototypa deteguntur. Jam superiore saeculo notatum est glossopetras esse dentes Lamiarum. Non ita pridem deprehensum accepi bufonium, quem vocant lapidem, esse piscis lupi. Qui baculi Sancti Pauli a Melitensibus appellantur, spinae sunt ad[57] histricibus marinis avulsae et saxo inclusae. Et oculos serpentum apud eosdem, dentes quorundam piscium esse breves et rotundos compertum est. Et quos illi serpentes

56. In Hannover Ms XXIII, 23b, pl. 4 appears here.
57. B: ab.

in diameter) from all other creatures found in the sea. But who has thoroughly explored the ocean's secret recesses and subterranean abysses? How many previously unknown animals did the New World give us? It is also conceivable that many animal species were transformed by these great upheavals. Lachmund displays our Ammon's horns in his *Fossilia patria*, whose images we have reproduced here.[58] But the careful English investigator of nature John Ray has examined these issues in greater detail.[59] Nor do I doubt that with such a great disturbance of things, the spoils of the sea were often carried from distant coasts; it is also well known that storms sometimes cast shells up onto the coasts of a kind that no fishermen haul out of nearby seas. When an immense mass of similar objects is assembled in one place, as in Malta, where one admires endless sea dog teeth, called glossopetrae, one attributes them not without reason to the whirlpools, where, after much agitation, those things are gathered and deposited in one place, on the basis of the combined action of movement and weight. Similarly, with crystallization, we see how different salts dissolved in the same liquid come together without confusion, like attracting like. Mostly, though, I think that the water, having found its way through narrow passages, often left behind what it was carrying. It thus necessarily came to pass that, given a flow constantly renewed and always pregnant with remains, an immense quantity of marine objects soon accumulated above the narrows.

[XXVII. *Glossopetrae, asterias, trochites, etc., are the remains of marine animals, and not games of nature*]

The true prototypes of these stones are finally being uncovered more and more through the observations of natural philosophers. Already in the last century, it was recognized that glossopetrae were sharks' teeth. I heard that the so-called toadstone was discovered to belong to the wolf fish somewhat later. What the Maltese call "St. Paul's sticks" are cast-off sea urchin spines enclosed in stone. Their "snake eyes" are the short, round teeth of some fish. And what they believe to be snakes, which

58. Lachmund 1669, 33–35. See fig. 4.
59. Ray 1692.

ipsos arbitrantur a D. Paulo in saxa versos, sunt vermes marini incrustati. Lapides Iudaici piciformes apud Bethleem a peregrinatoribus notantur. Valerius Cordus nostras meminit Iudaici lapidis vertebrae piscium similis. Lapis est Asterias dictus, stellae pentagonae specimen referens, quarum multae sibi impositae columellam striatam componunt. Et suspicatur Gassendus, ex spoliis atque incisuris qorundam[58] vermium formari. Nempe Trochitas quoque, et ex his conflatos Entrochos vertebris similes cum processibus seu apophysibus Lachmundus prope Hildesiam invenit, ubi et Cordus olim apud Agricolam vidit in commissuris marmoris e cinereo candidi, et in terra glutinosa inter Alfeldam et Einbeccam urbes. Sic lapis est Anglis a S. Cuthberto cognominatus caudam animalis referens ex plurimis articulis continuatam, qui ex insula affertur in littore Northumbriae, quam sanctam appellant. Sunt et quaedam hujus generis, quae ad corallii articulati ramos saxo incrustatos referuntur, de quibus omnibus extant eruditorum conjecturae, nosque ipsi satis credibile, ni fallor, fecimus, pisces in scissili lapide ex veris animalibus velut ectypos, superfuisse.

[XXVIII]

Et sane plerumque video, quantoquisque[59] in observando diligentior, et cum natura familiarior fuit, eo proniorem in nostram sententiam visum, ut peritissimi viri merito animalium exuvias, aut aliarum rerum reliquias putent obrutas, nec facile persuaderi sibi patiantur, organica corpora sine exemplo, sine usu, sine seminiis, praeter naturae consuetudinem in limo saxove, ineptis matricibus, nescio qua plastica facultate natas. Nam removere hinc oportet radiata quaedam corpora polygonorumque figuras regulares in crystallis, in granato, in reliquis gemmis et fluoribus, et variis mineris; tum in sexangula nive, in apum alveolis; in vitriolo etiam et alumine, et communi sale, nitroque et sale de cornu cervi et regulo martis stellato; caeteramque omnem naturae inanimae geometriam. Etenim haec externis appositionibus commode explicantur, ut in crystallismo;

58. B: quorumdam.
59. B: quanto quisque.

St. Paul turned to stone, are petrified marine worms.[60] The pilgrims find pine-shaped Jewstones[61] near Bethlehem. Valerius Cordus notes a Jewstone in our region which is like the backbone of a fish. There is another stone called *asterias*,[62] which displays a five-cornered star; when many of them are superimposed, they form a little grooved column. Gassendi suspected that they were formed out of the remains and fragments of certain worms. Lachmund also found trochites near Hildesheim, together with swollen entrochites made of them and looking like vertebrae, with their processes and apophyses. That is where, according to Agricola, Cordus once saw them in the cracks of ash-gray marble and also in claylike earth between the Alfeld and Einbeck. Similarly, the stone that the English named after St. Cuthbert, which comes from an island on the coast of Northumberland called the Holy Island, reveals the tail of an animal with many articulated segments that are joined together. There are also other things of this kind which belong to the petrified branches of articulated corals. The learned have conjectures about all of these things. But if I am not mistaken, we have made a plausible enough case that the fish in slate are the imprints of real animals.

[XXVIII. *But it is wrong to include the polygonal shapes that can be found in crystals among these*]

The more careful in observation and familiar with nature people are, the more they are inclined to adhere to our judgment, to the point that the most experienced people think, with good reason, that the remains of animals or of other things have indeed been buried: they are reluctant to believe that organic bodies, without precedent and without utility,

60. Marine fossils, found in abundance on the island of Malta, have long been associated with local legends, particularly with the traditional image of the St. Paul, who, bitten by a snake, turns it into stone as punishment.

61. *Lapides judaicos*, so named because of their abundance near Palestine, are the fossil spines of large sea urchins. When crushed into powder, they were used as medicine for the expulsion of kidney stones. See Gould 2002, 161–174. Leibniz here follows Agricola by interpreting this invertebrate fragment as the spine of a vertebrate, i.e., as "the backbone of a fish."

62. The term "asterias" does not refer here to starfish but to trochites, or "wheelstones." See glossary.

longeque ab operosiore illa structura animalium plantarumque absunt, quae post diligentem inquisitionem hactenus non nisi per similium semina, velut praeformata nasci constat, quantum hominibus observare datum est; explosa putredine prolifica et quicquid generationis aequivocae non barbare minus, quam falso memorabatur.

[XXIX]

Qui contra sentiunt, narratiunculis seducuntur, quae apud Kircherum quendam, aut Becherum, aliosque id genus credulos aut vanos scriptores de miris naturae lusibus et vi formatrice in magnam speciem verbis ornantur. Neque enim animalia tantum, et plantas, partesque eorum in saxis vident, sed et historias fabulasque[60] Christum et Mosen, in crusta Bumannianae specus; Apollinem cum Musis in Achate Pyrrhi; Papam et Lutherum, in Islebiensi petra; et solem cum luna stellisque in marmore.[61] Et meminimus vultus magnorum aliquot nostri temporis principum in una gemma ante aliquot annos passim monstratos ab Eilero Hamburgensi quodam ubi inprimis Alexandri VII. icon visenda extabat, et Reginae Christinae admirationi fuit. Quidam campos integros ostendunt tibiis gigantum stratos; alii aurum referunt ex vite Hungariae fruticescens, cum granis aureis intra uvas; et lapides passim celebrant cum tonitru dejectos, in clavae aut securis specimen formatos, credo quo melius ferirent. His adde quas vocant iridis scutellas, quasi a coelesti arcu lapsas vel excussas:

60. B: fabulasque: utpote qui.
61. B: in marmore deprehendisse sibi visi sunt.

without seeds and against the rules of nature, arose in the earth or in stone—as in some absurd womb—through some kind of plastic faculty. For one should exclude certain radiated bodies and the regular polygonal shapes of crystals, of garnets and other gems, of fluorspars, and of various ores; also hexangular snowflakes and beehives, vitriol, alum, common salt, saltpeter, hartshorn salt, star regulus of Mars, and all the other geometry of inanimate nature. For these things can be explained easily by external contiguity, as in crystallization. But they are far different from the more artful structures of animals and plants, about which one knows, after careful examination, that until now they have only arisen from the preformed seeds of similar beings, as far as human observation can ascertain. The notion of productive putrefaction and everything that was said, no less barbarous than false, about spontaneous generation have been overthrown.

[XXIX. *In which a certain lazy ingenuity, which invents things alien to truth, is rejected*]

Whoever believes the contrary is seduced by the fairy tales of Kircher or Becher, and of other credulous or vain writers of this sort, who describe the wonderful games of nature and its formative power, all embellished with a great display of words.[63] In fact, what they see in stones are not so much animals, plants, and parts of these, but fables, stories, and myths, such as Christ and Moses on the walls of the Baumann Cave; Apollo with the muses in the agate of Pyrrhus; the pope and Luther in the stone of Eisleben; and the sun, moon, and stars in marble.[64] We recall that a few years ago, a certain Eiler of Hamburg displayed the faces of some of the great princes of our time in a gem in which, when examined carefully, one could admire the inscribed portraits of Alexander VII and Queen Christine. Some people speak of entire plains spread with the leg bones of giants. Others report that gold sprouted from the vines in Hungary, with grains of gold among the grapes; and they often speak about stones that tumble down with thunder and are shaped like a club or an ax,

63. Kircher 1664, bk. 8, chaps. 8–9; Becher 1681, bk. 2.
64. Cf. Agricola 1546, bk. 6.

ficta pleraque aut semivisa, et illis similia, quibus Crollii imaginatio in rerum signaturis ludit. Quibus pictorem doctum oppono, qui nuper libello edito asseveravit, multa talia ostentata sibi, sed quanto attentius aspiceres, eo longius a similitudine abfuisse; cum contra in veris exuviis, quo scrutabere diligentius, eo manifestiora origines argumenta suppeditentur; totiesque deceptum se magnificis narrantium verbis non facile amplius fidem habere. In nostro monte S. Andreae, fodina Samsonis nomine, quae nunc quoque colitur, cruci fixum[62] Salvatorem cum spinea corona ex puro argento elaboratum exhibuisse fertur. Et alia fodina, cui nomen a Gratia Dei, ex argento hominem cunicularium praebuit habitu fossoris, portantem alveum metallo plenum, longitudine digiti. Sed haec imaginationis judicia sunt, non oculorum. Christianis et fossoribus facile occurrit animo, quod quotidie colunt, aut vident. Plerumque etiam ars adjuvat naturam, ne lepidae narrationis argumentum pereat. Adimitur aliquid ubi addi non potest. Et subtilius artificium circumforanei habent, quos *Mandragoram*, quam vocant, suam ita praeformare ex bryoniae radice scimus, ut crescendo ipsa sese in hominis speciem absolvat. Caeteroqui rudia passim lineamenta, quae casus formavit, supplet credulitas. Solent superstitiosae mulierculae certo anni die ovum exonerare in cyathum vitreum aqua plenum, vana spe divinandi. Quod si mox, ut saepe fit, velut domuncularum species, aut turrium in aqua apparent, res sibi lautas spondent. Ita si ligneae tabulae oleosum humorem illinas, videbis inaequali discursu varias rerum formas simulari; et, si duae tabulae politae subito invicem distrahantur, liquorem impositum velut flores delineasse. Sed sapienter Romanae eloquentiae princeps: Credo, inquit, non dissimilem rebus figuram aliquando apparere, at non talem, ut eam factam a Scopa diceres. Sic enim profecto se res habet, ut nunquam perfecte veritatem casus imitetur. Itaque ad genios superiores in formis saxorum jocantes cum doctissimo quodam viro (Conring) confugere, nihil necesse est.

62. B: affixum.

the better to strike, I believe.[65] To these you can add the so-called cups of Iris, which are supposed to have fallen, or been shaken out of, rainbows. Most of these are fictions or things half seen, and similar to the signatures of things, with which Crollius's imagination plays.[66] To these I oppose a learned painter, who declared in a recently published book that, though he had been shown many such things, the more carefully one observed them, the more tenuous the similarity.[67] With true remains, on the contrary, the more carefully and thoroughly one examines them, the clearer are the arguments furnished for their origin, even when one has been misled so many times by the magnificent words of those who tell these stories that it is not easy to retain one's faith. In our St. Andreasberg, in a mine by the name of Samson that is still in operation today, one is supposed to have discovered our Savior on the cross with a crown of thorns, fashioned of pure silver. Another mine, called God's Grace, yielded a finger-sized miner of silver, dressed in his miner's uniform and carrying a basket full of metal. But these are products of the imagination, not the eyes. For Christians and miners, what they worship or see everyday comes easily to the mind. Often, human art helps nature to concoct an appealing story. Something is taken away when it is not possible to add anything. We know that traveling merchants have an especially clever trick: they prepare their so-called mandragora from bryony root, so that it forms itself into the shape of a man as it grows. As for the rest, credulity fills in the rough outlines shaped by accident. Superstitious little women like to break an egg into a glass full of water on a certain day of the year, in the vain hope of divining the future. And if, as often happens, something like an image of little houses or towers soon appears in the water, they promise themselves riches. Similarly, if you smear a wooden tablet with oily liquid, you will see the shapes of various things in the uneven streaking of the oil. And if two polished tablets are suddenly separated from one another, the liquid placed on them will trace something like flowers. But as the prince of Roman eloquence wisely says, I think a form

65. "Thunder-stones," or *ceraunia*, were believed to have fallen from the sky, before they were identified as flint axes Cf. Laming-Emperaire 1965.

66. Oswaldus Crollius (1560–1608), physician and hermetic philosopher, wrote the *Treatise of Signatures of Internal Things*.

67. Agostino Scilla (1639–1700), painter, naturalist, and author of the widely read and translated book *La vana speculazione disingannata dal senso* (1670).

[XXX]

Quoniam autem Glossopetrae Luneburgicae inprimis celebrantur, post Melitenses, illustre Oceani terras operientis monumentum; non intempestive dabimus aliquid descriptioni earum, utemurque observationibus viri eruditi, qui peculiarem libellum impendit. Itaque prope Luneburgum ad radices montis, cui lateraria officina superstructa est, terra est salsuginosa sive aluminosa, nec pinguis adeo, qualis apud Georgium Agricolam describitur, sed macra, et tantum non alicubi sabulosa, nec ad lateres ducendos apta, nisi profundior actisque cuniculis effossa, donec pluviae solique exposita et humido probe irrigata lentescat.[63] Ibi ergo accidit subinde, ut a fossoribus in terrae visceribus meatibusque conchylia et unicornu, quod vocant, aut ebur et lignum fossile, turbines, brontiae, trochoides, ac Glossopetrae denique reperiantur. Agricola aliique eum secuti ex fodinis aluminis Luneburgenses Glossopetras petunt: sed has ager iste omnis ignorat; nec modum aluminis parandi novit, nec aut Corneri chronicon vetus, ineditum aut alia loci monumenta aluminis uspiam meminere. Et cum constet ejus coquendi artem vix trecentis abhinc annis a Rocca Syriae in Europam rediisse, (unde aluminis Roccae, non intellecta vulgo appellatio) atque in Italia primum exercitam serius in Germaniam penetrasse: satis manifestum arbitror, recentes satis (si quae fuissent) aluminis Luneburgici officinas esse debere, aut potius, quia res propinqua et celebris memoriam hominum, et annotantium diligentiam non efugisset,[64] nullas fuisse.

63. B: pluviae denique per aliquod tempus et soli exposita, et humido probe irrigata.

64. B: effugisset.

not dissimilar to real things sometimes appears, but never such that one could say it was produced by a Scopas.[68] And this is the way things are, that chance never perfectly imitates truth. This is why it is not at all necessary to seek refuge, as a certain learned man (Conring) does, in higher powers that play their jokes in the shapes of stones.

[xxx. *Where can the Lüneburg glossopetrae be found?*]

Since the glossopetrae of Lüneburg, along with those from Malta, are especially renowned for providing clear evidence that the ocean once covered the earth, it will not be inappropriate for us to offer a description of them, using the observations of a learned man who has produced a special pamphlet on the question.[69] Near Lüneburg, at the foot of a mountain upon which a brickworks has been built, the earth is salty or aluminous and not as rich as Georgius Agricola described it. Instead, it is thin, almost sandy, and thus unsuitable for making bricks, unless one digs the earth out of deeper shafts, exposes it to the rain and sun, and moistens it properly until it toughens. Thus, it often happens there that miners find shells, what they call unicorn or fossil ivory and wood, turbines, toadstones, trochoids, and finally glossopetrae in the viscera and veins of the earth. Agricola and those who follow him seek glossopetrae in Lüneburg's alum mines. [70] But that whole region lacks such mines. One does not know how to prepare alum there, nor do the old unpublished chronicles of Cornerus[71] or other records from this region mention alum anywhere. It is well known that the art of alum roasting came to Europe hardly three hundred years ago from Rocca in Syria (hence the name, *alumen rocae*, which is commonly misunderstood), and that it was first

---

68. Cicero, *De divinatione*, bk. 1. Scopas was a Greek sculptor of the late classical period.

69. In this reflection on glossopetrae, Leibniz follows a long tradition that started in the Italian Renaissance and culminated in Steno's demonstration (Steno 1667) that so-called petrified snake tongues were sharks' teeth (after he dissected the head of a shark). Steno's dissertation, little known in his lifetime, is viewed today as a classic of early paleontology. See Morello 1979.

70. Agricola 1546, bk. 6.

71. Hermann Körner, born 1402 in Lübeck.

Glossopetrae autem Luneburgenses nihil a Melitensibus differunt, nisi quod minores nostrae esse solent, nec ut illae in saxo, sed in terra. Linguam repraesentant non serpentis, ut vulgo volunt, sed pici potius, ut jam Agricola noster observavit. Lamiarum, cetacei generis piscium, aut canum marinorum dentes esse, vix jam amplius dubitatur, etsi aliter sentiat vir doctus, quem de his scripsisse dixi. Variant forma, ut in ipsis animalibus, nam serratae persaepe comparent in margine, interdum nudae. Color diversus; credo ab ambiente. Nostris fere nigricans aut subcineritius. Pars acuminata, quasi cornea, laevis et polita et dura est; contra postica et crassior ad radices, mollior est, spongiosa, scabra, obscuriorque. Et quemadmodum in his animalibus dentes plurimi incurvi sunt, atque introrsum versus gulam flexi, ita in Glossopetris, id est fossili dente, eadem figura apparet, ut dextra, an sinistra parte sederint, agnosci posse Scylla pictor notarit. Placet figuras subjicere, et nostrarum et Melitensium,[65] ut, qui Lamiarum dentes videre, ipso oculorum judicio, quam nihil intersit, arbitrentur. Caput etiam canis carchariae ex Stenonis delineatione cum dentibus subjiciemus in comparationem.[66] Nec mirum esse debet, quod maxillae ipsae non comparent, quales interdum terrestrium animalium ex Baumanni antro aut Scharzfeldensi specu eruuntur, cum dentibus infixis. Nam dudum observatum est a curiosis, Lamiarum dentes non aeque in ore firmos esse, sed membranae tantum haerere. Itaque evulsi motu aquarum, longiusque provecti, maxillas suas facile deseruere. Praeterea pronum est, credere, etiamsi una mansissent, maxillam piscis consumtam tempore, aut vi ambientis. Nam et in sepulcris constat, dentes prae caeteris animalis partibus inprimis aevum ferre. Interim ossa marinarum

65. In Hannover Ms XXIII, 23b, pl. 6 appears here.
66. In Hannover Ms XXIII, 23b, pl. 7 appears here.

practiced in Italy and entered Germany later. I am therefore quite certain that any alum workshops in Lüneburg, if there ever were any, would have to be very recent, or, what is even more likely, that there never were any, since something so prominent and so recent would not have escaped human memory and the careful attention of chroniclers.

[XXXI. *Glossopetrae are sharks' teeth*]

The Lüneburg glossopetrae do not differ from the Maltese, except that ours are usually smaller and, unlike the latter, are not found in stones but in the ground. They do not look like a snake's tongue, as the common people claim, but rather like that of a woodpecker, as our Agricola has already noted.[72] One hardly doubts any more that they are teeth from some kind of whalefish or from sea dogs, even if the learned man—the one who, as I said, has written about them—sees it differently. Their shapes vary, just as with the animals themselves, for the edges of the glossopetrae are often jagged and sometimes smooth. Their color varies, which stems, I think, from their surroundings. Ours are often blackish or almost gray. The pointed end is smooth and hard, like horn; but the back end, which is thicker toward the root, is softer, spongy, rough and darker. And just as these animals have mostly curved teeth that are turned toward the inside of their mouths, so it is with glossopetrae, that is, with fossil teeth. It is therefore possible, as the painter Scilla noted, to recognize whether they sat on the right or on the left. I would like to append images of our glossopetrae and of the Maltese, so that whoever has seen shark teeth can testify as an eyewitness that there is no difference.[73] I would also like to append, by way of comparison, the head of a great shark with its teeth, from a drawing by Steno [fig. 7]. One should not wonder that the jaws themselves are unlike those of land animals, with the teeth still planted in them, that one sometimes digs out of the Baumann or Sharzfeld caves. For it has long since been observed by the inquisitive that sharks' teeth are not firmly fixed in their mouths, but

72. Agricola 1546, bk. 6.
73. See fig. 6.

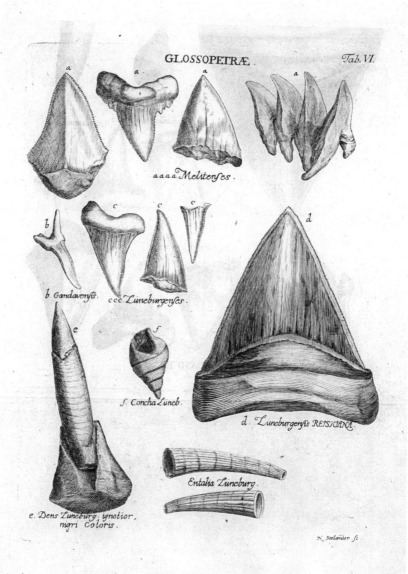

GLOSSOPETRÆ.

Tab. VI.

a a a a Melitenses.

b. Gandavensis.   *c c c Luneburgenses.*

e.   *f.*

f. Concha Luneb.

d.

d. Luneburgensis REISKIANA.

Entalia Luneburg.

*e. Dens Luneburg. ignotior,
nigri Coloris.*

N. Seelander fc

FIGURE 6. *Glossopetrae*

When he dissected the head of a shark in 1666, Nicolaus Steno demonstrated that glossopetrae, traditionally regarded as petrified snakes' tongues, were sharks' teeth. In *Protogaea*, Leibniz referred to Steno's "A Carcharadon-Head Dissected" (1669), in which this discovery was communicated (*Protogaea*, §XXX). The issue figured prominently, because glossopetrae were often found in landlocked places like Lüneburg, thus providing evidence that the ocean had once covered the earth (*Protogaea*, §XXXI).

This figure (plate 6 in Scheidt's 1749 edition), signed by the engraver Nicolaus Seelander, depicts glossopetrae from various places—Malta (*a*), Ghent (*b*), and Lüneburg (*c*)—with engravings based on the original specimens. It also displays other specimens with some similarity to glossopetrae, such as "(*e*), a more unusual tooth from Lüneburg, of dark color," and "(*f*), a shell from Lüneburg; and tube shells (Entalia) from Lüneburg."

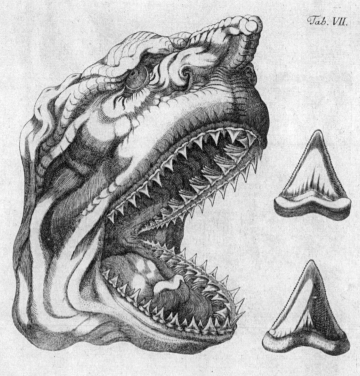

FIGURE 7. *The Great Shark*

Leibniz chose this illustration for *Protogaea,* even though it was quite fanciful. The image, which appeared in Nicolaus Steno's "A Carcharadon-Head Dissected" (1669), suggests the importance of Steno's work for Leibniz, demonstrating the organic origin of glossopetrae by representing them both inside the shark's mouth and outside it. Leibniz pointed out that, unlike the teeth of land animals, shark's teeth were "not firmly fixed in their mouths" (*Protogaea,* §XXXI). This explained why individual glossopetrae were often found.

Steno himself borrowed this image (plate 7 in Scheidt's 1749 edition) from a sixteenth-century manuscript by Michele Mercati, which appeared later under the title *Metallotheca vaticana* (1717). Nicolaus Seelander made the final engraving, based on Steno.

beluarum cum dentibus piscium in Luneburgensi agro aliquando fuisse effossa Agricola in literas[67] retulit.

[XXXII]

Porro Glossopetrae magni in Medicina usus habentur, quem non tantum veteres praedicarunt, sed et Melitenses publicis programmatibus venditant, et D. Paulo adscribunt, qui serpentum non exarmarit tantum vim noxiam, sed et in salutem mortalium verterit, linguis in lapidem alexipharmacum mutatis. Inde passim reperias, auro argentoque inclusam hanc, ut ita dicam, infimam gemmarum, sive ut collo appendatur, contra fascinationes, nescio quas, sive ut a[68] scypho insertis antidotum bibatur contra venena. Ita scilicet facti homines, ut quicquid specie aliqua praestat, etiam virtute eminere arbitrentur; communi errore in naturalium rerum arbitrio civiliumque. Inde tot de viribus gemmarum narrationes, et materia Medica fabulis inflata. Glossopetras tamen non a terra tantum ambiente, sed et per se posse arbitror, quod cornu cervi, quod gammarorum lapilli oculiformes, quod ebur fossile, et dentes ex Scharzfeldensi specu eruti, totaque Germania expetiti in Medicinam. Scilicet non tantum colicae et faucium erosioni, pustulisque ab acri humore excitatis mederi, sed et intus acidum, naturae inimicum, aggredi ac sorbere, atque ita liberatis spiritibus movere sudorem deprehenduntur. Unde vis quaedam medica,[69] quae deinde a credulis usque adeo in majus attollitur, ut jam quae malignitati lentae restistunt, etiam praesentissimum venenum in poculo obtundere aut prodere credantur. Ex omnibus autem Glossopetrae usibus nullum certiorem arbitror, quam ad dentifricia, quod ex contusis pulvis duritie sua, et quadam asperitate commendetur, ac dens denti minus noxius videatur.

67. B: litteras, belluarum.
68. B: ut ab illis.
69. B: vis quaedam Medica iis negari nequit.

only hanging by the skin.[74] Thus, torn out by the motion of the water and carried far away, they easily become separated from their jaws. Besides, it is easy to believe that the jaws would have been devoured by time or by the force of their environs, even if they had remained whole. For it is also well known that, in graves, teeth last much longer than all other animal parts. Still, Agricola reported in his letters that the bones of sea monsters with fish teeth had been dug up in the area around Lüneburg.

[XXXII. *The medical use of glossopetrae*]

Moreover glossopetrae have great use in medicine, which not only was proclaimed by the ancients but is also advertised by the Maltese in published pamphlets.[75] They attribute it to St. Paul, who not only destroyed the harmful power of the snakes but even turned it to the benefit of mortals, by changing snakes' tongues into a stone with healing powers. Hence, one often finds this lowest of gemstones, as I would like to call it, encased in gold or silver to be hung around one's neck against all kinds of magic arts, or placed in a cup as an antidote to poisons. For it is human nature to believe that if something looks remarkable, it possesses excellence as well—a common error in matters of nature and of state. From this come many stories about the powers of gemstones and also *materia medica* exaggerated by fables. I nevertheless believe that glossopetrae have the same capacity, coming not only from the surrounding soil but also from themselves, as hartshorn, the eye-shaped stones of crabs, fossil ivory, and the teeth excavated from Sharzfeld Cave, which are demanded for medicine in all of Germany. They not only heal colic, sore throat and the little blisters caused by sharp juices; they also attack and absorb the inner acid, which is hostile to nature, and promote sweat by liberating breath. From this stems a certain healing power, which has been so ex-

74. In this section Leibniz uses various terms for sharks, including *canis marinus* (sea dog), *canis carcharias*, and *lamia*. The term "sea dog" was used to denote smaller sharks, while *canis carcharias* and *lamia* referred to larger animals like the great white.

75. As petrified snakes' tongues, glossopetrae were considered antidotes to poisons and animal bites, a belief justified by the Maltese legend of St. Paul, who was said to have trasformed snakes into stone. Later identified as shark teeth, they could be used, as Leibniz claimed, to clean teeth, a medical interpretation that still relies on the doctrine of analogy or signatures.

Belemnitae sunt lapides a jaculo dicti; dactyli etiam idaei appellantur;
Saxonibus composita voce ab Ephialte et sagitta, Alpschos, quod contra
noctis ludibria et oppressiones valere dicantur. Medici quidam Lyncurium nominant, praesertim cum pellucet. Et jam diximus, passim prope
Hildesiam et prope Neostadium ad Leinam inveniri. Figura est sagittae,
nam ex ampla radice plerumque in aciem desinunt; fere omnibus a natura
inest quaedam quasi rima, qua fit, ut facilius in longitudinem diffindan-
tur. Paleas ut succinum allicit, intus terram, arenam aut alium lapidem
claudit. Nitet plerumque ut cornu, et saepe cornu ustum olet, ut suspicio
sit, intervenisse aliquid ex animali regno.[71] De cornu Hammonis dictum
est supra. De Osteocolla, quae Germanis Ossifraga dicitur, Beinbruch, et
ossis specie cum inclusa velut medulla reperitur prope Alfeldam, com-
perendinare adhuc malim. In arena reperitur forma interdum corallii,
crassitie brachii.

Sed quia generatim de marinis exuviis lapide inclusis diximus, et
Glossopetram, id est Carchariae dentem peculiariter exposuimus,
placet et lapides nostri tractus conchyliis foetos exhibere, ex[72] diligen-
tis in vicinia Medici, cui Oryctographiam Hildesheimensem debemus,
observatis:

(p. 40 Strombites etc. ad p. 41 Lachmanni)

70. See "Introduction," xiii, and the appendix.
71. In Hannover Ms XXIII, 23b, pl. 8 appears here.
72. B: ex Friderici Lachmanni.

aggerated by the credulous that they believe what can withstand a slow malady would also weaken or reveal the strongest poison in a cup. But of all the uses of glossopetrae, I believe none is more reliable than for cleaning teeth; the powder from crushed teeth is recommended because of a certain hardness and roughness, and because tooth against tooth seems the least harmful.

[XXXIII. *Belemnites, osteocolla, shell-filled stones, and fossil ivory*]

Belemnites[76] are stones named after the javelin.[77] They are also called "fingers of Ida."[78] The Saxons call them "Alpschos," a word combining nightmare and arrow, because they are said to be powerful against the illusions and oppressions of the night. Some doctors call them "lynx stones," especially when they are transparent.[79] One finds them, as we already said, here and there around Hildesheim and near Neustadt on the Leine River. They have the shape of an arrow, mostly with a wide root ending in a point. Almost all of them have by nature a kind of inner cleft, which makes it easy to split them lengthwise. Belemnite attracts straw, like amber; internally it encloses earth, sand, or another stone in itself. It usually glistens like horn and often smells like burnt horn, so one might suspect that something from the animal kingdom has

76. In petrifactions, the suffix "-ite" indicated something stony, while the root of the word denoted what the object resembled (without asserting anything about its actual essence). Belemnites resemble javelins, trochites look like wheels, ctenites resemble combs, etc. Leibniz, however, goes beyond resemblance in attempting to identify the actual origin of these different objects.

77. From the Greek, βέλεμνον, or "dart."

78. In English, they were often just called "finger stones."

79. Belemnites, smooth cylindrical and oblong fossils, were also called "lynx stones," since they were viewed as coagulated lynx urine. According to the doctrine of analogy, they were considered helpful in breaking up kidney stones and were believed to help against nightmares and witchcraft. These objects puzzled naturalists until they were recognized in the nineteenth century as the skeletons of an extinct species of cephalopod resembling a squid.

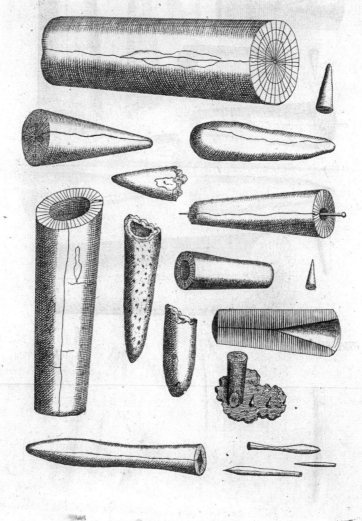

Tab. VIII.

FIGURE 8. *Belemnites*

This engraving presents a series of belemnites, or "lynx stones." Until the nine-teenth century, when they were recognized as the internal skeletons of extinct cephalopods, belemnites puzzled naturalists. (See Huxley 1880.) Leibniz guessed that these "fingers of Ida" came from the animal kingdom, on the basis of some of their physical properties (*Protogaea*, §XXXIII).

This unsigned engraving (plate 8 in Scheidt's 1749 edition) was modeled on the sketches in Lachmund's *Oryktographia* (Lachmund 1669, 26–28). See the appendix.

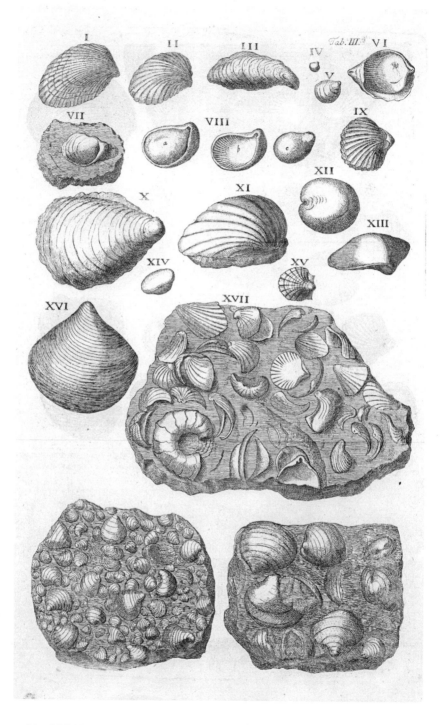

FIGURE 9. *Shells in Stone*

This engraving depicts a number of shell-filled stones and other petrifications. It includes images of stony conchites, myites, and ostracites (that is, petrified, trumpetlike shells, mussels, and oysters). The plate was reproduced from Lachmund (1669), whom Leibniz described as "a diligent physician from the region, to whom we owe the *Hildesheim Oryktographia*" (*Protogaea*, §XXXIII). It provided a visual demonstration of the large number of seashells present in and around Hildesheim, a town many miles from the sea.

The unsigned engraving (plate 3 in Scheidt's 1749 edition) was modeled on sketches from *Oryktographia* (Lachmund 1669, 43–46). Lachmund provided a detailed key to the engraving (see the appendix).

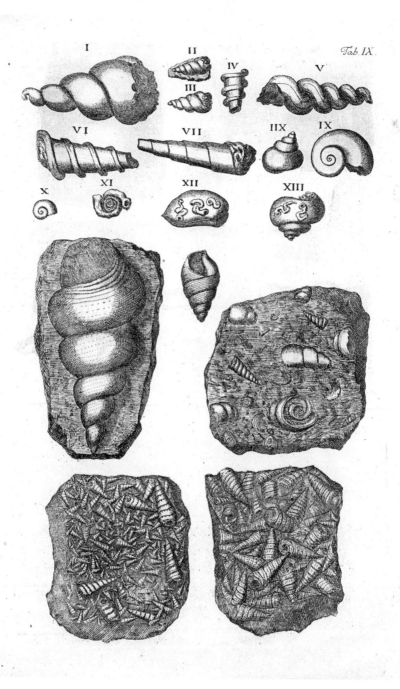

FIGURE 10. *Strombites*

This engraving depicts a wide array of "strombites," that is, a kind of petrified aquatic snail. As Lachmund explained, "it runs from wide to thin and ends in a spiral wound from the right. Sometimes it is short, sometimes nine inches long. . . . It is found in the quarries of Galgenberg, and in a new part of the city when digging the cellars where wine and beer are usually kept" (Lachmund 1669, 40). Lachmund's plate, later reproduced in *Protogaea,* showed a wide variety of shells collected within a single stony matrix, demonstrating once again that they "did not arise in the rock, but were mixed together with the mud" (*Protogaea,* §XXIV).

This unsigned engraving (plate 9 in Scheidt's 1749 edition) is modeled on sketches from the *Oryktographia* (Lachmund 1669, 45–48). For Lachmund's original key to the images, see the appendix.

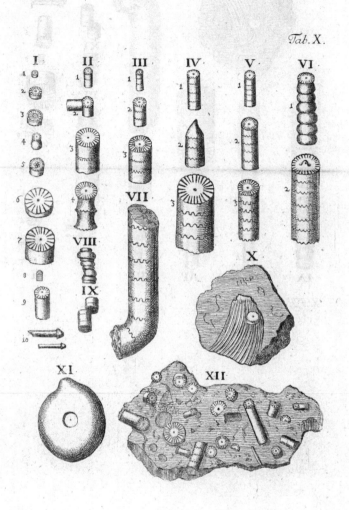

Tab. X.

FIGURE 11. *Trochites*

This engraving shows a selection of trochites, later identified as the joints of fossil crinoids. As Lachmund noted, the trochite has "the shape of a wheel; its round part is smooth, and from the center of its cross section spokes reach toward the outside of the disk, just like those of a wheel, and they jut out so much that grooves are created" (Lachmund 1669, 52).

This unsigned engraving (plate 10 in Scheidt's 1749 edition) was modeled on sketches from the *Oryktographia* (Lachmund 1669, 53–55). For Lachmund's original key to the images, see the appendix.

Sequuntur figurae cum adjecta explicatione qualicunque:[73]

(I. Conchites cinereus striatus etc. p. 41. ad p. 50 fin.)

Quanquam autem de Trochitis et multos Trochitas continente Entrocho aliquid jam ante attigerimus, placet tamen hic oculis nostrates subjicere.

(a p. 52 ad p. 56)

Ebur quoque fossile eodem, quo Glossopetrae aliaque marina, loco prope Luneburgum erui dictum est. Ac de ebore quidem suspicio venit, aliquando non tam ex Elephanti cornu esse, quam a Rosmari dente. Equi scilicet marini, aut similis de Phocarum ingentium genere animalis (*Walrossen*), quorum greges in Oceano Septentrionali piscatoribus balaenarum occurrunt, et parati ex dentibus capuli eburnis etiam praestantiores, et candoris servantiores censentur. Balaenae vertebram petrificatam in Musaeo habuit laudatissimus Lachmundus, Medicus Hildesiensis.

[XXXIV]

Conringius mihi testis erit, ex specubus nostris celeberrimis, Baumanniano et Scharzfeldensi marinarum beluarum,[74] aliorumque ignoti orbis animalium ossa integra frequenter erui, neutiquam ibi nata, sed translata illuc ex Oceano, violentia aquarum. Et vero possim ego, inquit, mon-

73. In Hannover Ms XXIII, 23b, pls. 3, 9, and 10 appear in the insterted Lachmund text that follows this phrase. See the appendix.

74. B: belluarum.

intervened [fig 8].[80] Ammon's horn was spoken of above. I would prefer to withhold judgment about osteocolla, which the Germans call *Beinbruch*, and which one finds near Alfeld in the form of bones that contain something like marrow. One sometimes finds them in the sand, coral-shaped and as thick as an arm.

But since we have spoken in general about seashells enclosed in stone, and since we explained glossopetrae (that is, sharks' teeth) in particular, I would also like to describe the shell-filled stones of our region on the basis of the observations of a diligent physician from our neighborhood, to whom we owe the *Hildesheim Oryktographia*:

(see p. 40, strombites, etc., to p. 41 in Lachmund)[81]

There follow the figures together with explanations:

(I. Conchites cinereus striatus, etc., on p. 41 to the end of p. 50)[82]

Although we already said something earlier about trochites, and about entrochites that contain several trochites, I would nonetheless like to offer those of our region here for observation.

(from p. 52 to p. 56)[83]

It is also said that fossil ivory is extracted near Lüneburg, from the same place as glossopetrae and other marine bodies. But some doubts have been raised regarding ivory, suggesting that it may sometimes come not from elephants but from rosmarian teeth. For it is clear that sword hilts made of ivory teeth from the marine horse or a similar animal, a kind of very large seal (*Walrossen*), schools of which are seen by the whale hunters in the Northern Ocean, are valued because they are even more prestigious and keep their white color better. The very esteemed Lachmund, physician at Hildesheim, had the petrified vertebra of a whale in his museum.

[XXXIV. *Bones, jaws, skulls, and teeth found in our region*]

Conring[84] will be my witness that entire bones of sea monsters and of other animals from an unknown world are often excavated from our

80. Cf. Agricola [1546] 1955, bk. 5.

81. Lachmund 1669.

82. See figs. 5, 9, and 10. For Lachmund's accompanying text, see the appendix.

83. See fig. 11. For Lachmund's accompanying text, see the appendix.

84. Hermann Conring (1606–1681), professor of law in Helmstedt, who also made significant contributions to politics and medicine.

strare volentibus et crania et maxillas inferiores belluarum cum insertis dentibus, ut penitus tollatur error de dentibus ex terra natis. Addit talia ossa interdum et rapida vi Leinae fluvii egesta esse. Itaque nihil prohibet, peregrina animalia ad nos undarum vi advecta esse, quanquam Elephanti dentibus minus fidam; nam ad Rosmarum referri posse, paulo ante judicavi. Quales forte fuerint, quos apud Mexicum effossos tradunt, cum nulli hodie in America Elephanti deprehendantur. Idem de ponderosis illis dentibus dixerim, qui ad Elephantum referuntur, Moschis Mammotekoos dici, quasi beluarum[75] ossa, Witsenius tradit. Nec tamen obstinate obnuerim, vera Elephantorum ossa reperiri, certe velut partem tibiae ex Scharzfelsae antro vidi.[76] Sive olim latius sparsa per orbem fuerint haec animalia, quam hodie, mutata aut ipsorum aut soli natura, sive aquarum impetu longissime a patria abrepta credamus. Et vero Phocarum et Narwallorum greges stabulari simul constat in cavitatibus litoralium scopulorum, qualia esse potuerunt haec antra, quo tempore huc Oceanus perveniebat. Cum vero terrestrium quoque animantium spolia deprehendantur, malim Conringio assentiente vastae inundationes colluviem credere, quando aquae per angusta foramina et specuum exitus aditum in subterranea reperientes, quae advehebant, deseruere in vestibulis. Verisimillimum enim est, orbem aquis mersum naturaliter detegi non potuisse, nisi magna parte humoris subterraneis locis recepta; qualis erat in Hierapoli Syriae hiatus immensae profunditatis apud Lucianum de Dea Syria, in quem stato anni die vicinae gentes certatim aquam quasi impleturae frustra fundebant. In quo diluvii aquas absorptas narrabant Sacerdotes, et populi credebant.

75. B: belluarum.

76. B: Certe dentes et velut partem tibiae aliaque ossa ex Scharzfeldensi antro vidi, quae alius animalis, quam Elephanti esse nemo dixerit, sive olim latius sparsa per orbem fuerint haec animalia.

most famous caves: Baumann and Scharzfeld. They did not arise in those places but came from the ocean, carried there by the violence of the waters. I could, so he says, show anyone who wishes the skulls and lower jaws of huge animals, with teeth still in them, in order to refute the false notion that teeth are born in the earth. He adds that such bones were also sometimes ejected by the swift force of the Leine River. Therefore nothing keeps us from supposing that these foreign animals could have been brought to us by the force of the waters. I certainly have less faith in the elephant teeth; that they could be attributed to a walrus, I just stated above. The teeth that reportedly have been unearthed in Mexico are probably of the same kind, since no elephants are found in America today. I would also say the same about those massive teeth that are attributed to elephants, and that the Muscovites call *Mammotekoos* as the bones of large animals, as Witsen reports.[85] Nevertheless, I will not completely deny that real elephant bones are found. I in any case, saw something like part of a shinbone dug out of Scharzfeld Cave. We can assume either that these animals were once more widely distributed across the globe than today, because their own nature or the nature of the ground has changed, or that they have been carried very far from their homeland by the strength of the waters. It is well known that schools of seals and narwhals gather together in the caves of seaside cliffs, like these caves also might have been at that time, when the ocean stretched to here. Since, however, the vestiges of terrestrial animals are also found, I prefer to believe, with Conring, that they were brought together by a great flood, so that when the waters found an entrance to the underworld through narrow holes and openings of caves, they deposited what they were carrying at the gate. It is in fact very likely that the earth, completely submerged underwater, could not have reemerged naturally unless a great part of the water was absorbed by underground spaces. Of this kind was the incredibly deep fissure in Syrian Hierapolis, mentioned by Lucian in *de dea Syria*; on a certain day of the year, the neighboring peoples competed, pouring water in as if they wanted to fill it.[86] The priests said that this abyss had swallowed the waters of the Great Flood, and the people believed it.

85. Nicolas Witsen, a Dutch magistrate and traveler who possessed a rich collection of antiquities and curiosities, was the first in Western Europe to report about the mammoth from northern Asia in his travel narrative *Noord en Oost Tartaryen* (1692).
86. Cf. Burnet 1681, bk. 1, chap. 7.

XXXV

Cum cornua Monocerotis, quibus passim superbiebant olim conditoria rerum peregrinarum, et nunc quoque plebeii oculi in stuporem dantur, a piscibus septentrionalis Oceani esse demonstrarit Bartolinus, credere fas est, unicornu fossile, quod nostrae quoque regiones praebent, ejusdem originis fuisse. Dissimulare tamen non oportet, Monocerotem quadrupedem equi magnitudine reperiri apud Abyssinos, si credimus Hieronymo Lupo et Balthasari Tellesio, Lusitanis. Terestris quoque animalis speciem magis referebat sceleton, in vicini nobis Quedlinburgi monte Zeunickenberga intra rupem anno saeculi sexagesimo tertio detectum, cum calcis materia effoderetur. Testis rei est Otto Gerikius, Magdeburgensis consul, qui nostram aetatem novis inventis illustravit, primusque mortalium antliam reperit, per quam vasis aer educitur, miraque spectacula ab inventore in Comitiis Ratisbonensibus anni 1653. coram Caesare edita sunt; qua deinde a Roberto quoque Boilio, Anglo, Corkiorum in Hibernia Comitis fratre, pro summo viri ingenio mirifice exculta, novo experimentorum thesauro locupletati sumus. Gerikius igitur libro de vacuo edito, per occasionem narrat, repertum sceleton unicornis in posteriore corporis parte, ut bruta solent, reclinatum, capite vero sursum levato, ante frontem gerens longe extensum cornu quinque fere ulnarum, crassitie cruris humani, sed proportione quadam decrescens. Ignorantia fossorum contritum particulatimque extractum est, postremo cornu cum capite et aliquibus costis, et spina dorsi atque ossibus Principi Abbatissae loci allata fuere. Eadem ad me perscripta sunt; additaque est figura, quam subjicere non alienum erit.[77]

77. In Hannover Ms XXIII, 23b, pl. 12 appears here.

[XXXV. *The unicorn's horn, and an enormous animal unearthed in Quedlinburg*]

Bartolinius has demonstrated that the horns of the unicorn, which were in the past the most celebrated ornaments in the displays of cabinets of curiosities, and which today still amaze the eye of the crowd, come from fish of the Polar Sea.[87] It is right to believe that fossil unicorn, which also appears in our region, was of the same origin. Nevertheless, we should not disguise the fact that a four-footed unicorn the size of a horse has been found in Abyssinia, if we can believe the Portuguese Hieronymus Lupus and Balthazar Tellesius.[88] The skeleton that was found in 1663 near Quedlinburg on the Zeunickenberg in the rock, while lime was being excavated, also looked more like a land animal. Otto Guericke, the mayor of Magdeburg, who has enriched our age with new inventions, and who was the first of all mortals to discover a pump that removes the air from a vessel, witnessed the thing.[89] In 1653, this inventor conducted wonderful demonstrations at the Reichstag in Regensburg in front of the emperor; then, to the advantage of this man's genius, these demonstrations were wonderfully improved by the Englishman Robert Boyle, brother to the count of Cork in Ireland, so that we have been enriched by a new treasure of experiments. In his book about the vacuum, Guericke mentions in passing that the skeleton of a unicorn was found with the rear part of its body bent back, as is common with animals, but with a raised head and carrying on its forehead an extended horn about five yards long; the horn was the width of a human leg and tapered gradually. Because of the ignorance of the workers, it was broken and brought out in pieces. Eventually, the horn, together with the head, several ribs, dorsal vertebra, and bones were brought to the town's serene abbess. The same thing was reported to me. An illustration was included, which it will not be inappropriate to append here [fig. 12].

87. See Bartholin 1645. On the myth of the unicorn in early paleontological literature, see Cohen 2002.

88. Balthazar Tellesius (b. 1595), a Portugese historian and Jesuit, wrote a collection entitled *General History of Ethiopia* (1660), which was later translated and shortened by Jean de Thevenot.

89. Leibniz relies entirely on Guericke's narrative, from which he quotes extensively. Cf. Guericke 1672.

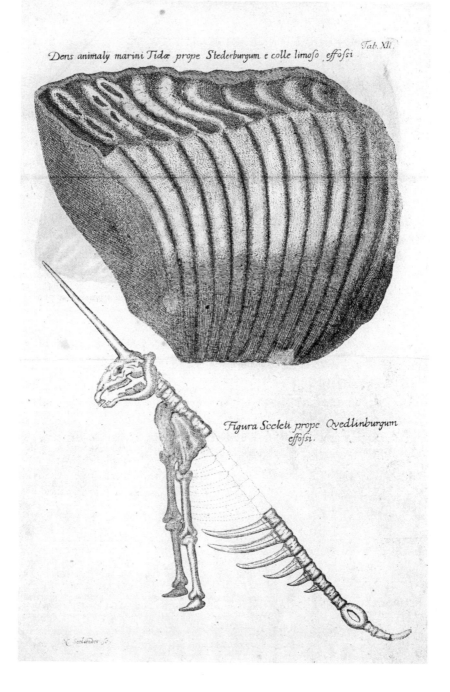

Dens animalij marini Tidæ prope Stederburgum e colle limoso effossi

Tab. XII.

Figura Sceleti prope Qvedlinburgum effossi.

T. Schlender sc.

FIGURE 12. *The Unicorn*

The remains of a "unicorn" were discovered in a gypsum quarry near Quedlinburg in 1663. Leibniz refers to the account provided in Otto von Guericke's 1672 *Experimenta nova (ut vocantur) Magdeburgica de vacuo spatio.* Paleontologist Othenio Abel called the illustration, which includes the image of a mammoth molar tooth (*top*) and a collection of bones (*bottom*), the first attempt to reconstruct a vertebrate in the history of paleontology (Abel 1925). Note the dotted lines, which hypothetically restore the missing parts of the animal in this odd skeletal reconstruction. Leibniz describes Guericke's discovery in §XXXV of *Protogaea.*

Nicolaus Seelander made the engraving (plate 12 in Scheidt's 1749 edition), but its source remains a mystery. The image here does not appear in Guericke's book. Leibniz probably had his engraver reproduce, and improve upon, an existing drawing that was circulating in contemporary periodicals. The caption at the top reads, "Tooth of a marine animal unearthed from a hill of clay at Tidae, near Stederburg." The lower caption reads, "Image of a skeleton dug up near Quedlinburg."

XXXVI

Sed res admonet, ut de specubus nostris distinctius dicam, nam ambas ipse sum ingressus, nimirum in Ducatus Grubenhagii extremo, qua ex Brunsvicensibus ad Thuringos tenditur, in monte arx est Scharzfelda Comitibus propriis olim habitata, historiae nostrae memorandis. Collis assurgit in vicinia, ubi vivo saxo velut templum incisum. Inter collem et arcem alius stat collis paulo minor, in quo antrum est, quod incolae a nanis appellant. Credo quod homini non pygmaeo intus rependum est: Et vulgi fabulis, cantionibusque antiquis, memorantur homuncunculi,[78] inaccessis montium cavernis habitantes, immensarum opum custodes, qualis Nibelungus nominatur in carmine, quo Sifridus Corneus, Wangionum Regis, si credimus, filius, celebratur. Foramen est in latere collis, qua villa Scharzfeld respicit, quo si oculos vertas, dextra Hertzbergam, laeva Scharzfeldam arcem habebis, et duo longe dissita castella prospicies in montibus gemellis, qui Germanis dicuntur Gleichen, id est pares. Porro antri ingressus est ulnarum circiter quinque altitudine, latitudine trium et semis. Per 15. pedes recta descenditur, ibi velut atrium est, quod porro in montem ducit. Limo nigricante vel fusco infectum est solum, et in eo paululum progressis foramen occurrit ex superiore loco aerem admittens. Sexaginta ultra passibus manente latitudine contrahitur altitudo, ubi pronum incedere opportet:[79] Inde rursus aperitur, mox denuo contrahitur locus. Postremo ad angustias ventum est, ubi velut stiriae concretae; sed longius repere non placuit. Nihil enim ultra magnopere visendum offerri aiebant ductores, sed sub ampliore fornice ad rotundum foramen perveniri, per quod in laxius iterum spatium aditus detur, ibi specus terminum esse. Porro in toto antro multa sunt saxorum fragmina tenui crusta obducta. Si fodias sub primo limo occurrit marga, in mollem lapidem indurata octonum aut duodenum pollicum strato. Subtus terra est nigra, plenaque non tantum fragminibus margae at[80] fornicis, sed et multis animalium ossibus, ruptis quidem fere aut disjectis, sed ut partem corporis facile distinguas. Dentes quoque multiplices varii coloris,

78. B: homunculi.
79. B: oportet.
80. B: ac.

[XXXVI. *Scharzfeld Cave and the bones that have been found in it*]

But it is time that I speak in more detail about our caves, for I have visited both of them. On a mountain at the border of the Duchy of Grubenhagen, which stretches from Brunswick to Thuringia, sits Scharzfeld Castle, once occupied by its own counts, of whom our history will speak. Nearby there rises a hill, upon which something like a temple is cut into the natural rock. Between this hill and the castle stands a somewhat smaller hill, in which there is a cave that the locals have named after dwarves, probably, I think, because someone who is not a dwarf must crawl in it. Folktales and old songs mention homunculi who live in inaccessible mountain caves and guard great riches, like the Nibelung named in the poem that sings of Siegfried the Horned, a son of the king of the Vangionen, if we believe it.[90] The cave's entrance is on the side of the hill that faces the town of Scharzfeld. From there, turning your eyes, Herzberg is to the right, on the left you have Scharzfeld Castle, and farther in the distance you see two castles on twin mountains, which in German are called Gleichen, because they are the same. The entrance to the cave is approximately five yards high and three and a half wide. One goes straight down fifteen feet, where there is a kind of atrium that leads farther into the mountain. The ground is covered with black or dark brown mud. Moving a little farther along, one encounters a hole through which air enters from above. After sixty steps the width remains the same and the height decreases, so that it is necessary to proceed bent over. Then it opens up but soon constricts again. Finally comes a narrow passage, where there was something like hardened icicles. But I did not want to go any farther, because the guides said that there was not much more to see: passing through a higher vault, one would come to a round hole that again offered access to a roomier space, where the cave ends. Throughout the whole cave there are many pieces of rock covered by a thin crust. If you dig through the initial mud, you encounter marl that has hardened into a layer of soft stone eight or twelve inches thick. Underneath it, the earth is black and filled not only with pieces of marl and cave vault, but also with many animal bones. These are

90. "With his hand he slew a dragon, and bathed him in its blood, that his skin is as horn." *Nibelungenlied* 1999, adventure 3. Cf. Kircher 1664, bk. 8, chap. 4.

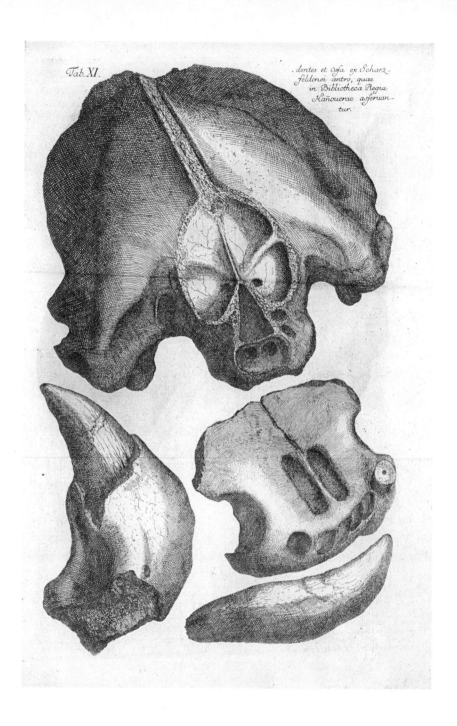

Tab. XI.

.dentes et ossa ex Scharz-feldensi antro, quae in Bibliotheca Regia Hanouerae asseruantur.

FIGURE 13. *Bones and Teeth from Scharzfeld Cave*

This engraving shows vertebrate fossil remains—skull parts and two canines, probably those of a cave bear. Because of their value as medicines, these remains were widely collected from caves, along with so-called unicorn horns, that is, ivory from the fossilized tusks of Proboscidea (e.g., elephants). These were discovered in Scharzfeld Cave, which Leibniz described this way: "The earth is black and filled not only with pieces of marl and cave vault, but also with many animal bones. These are indeed broken in pieces or scattered about, but you can still easily distinguish the body parts. There are many kinds of teeth of various colors; they are often shiny and, not infrequently, are still inserted into pieces of jawbone. Some are so large that they cannot be ascribed to any animal known to us" (*Protogaea*, §XXXVI).

This unsigned engraving (plate 11 in Scheidt's 1749 edition) was modeled on specimens from the Royal Library in Hannover. The original caption at the top reads, "Teeth and bones from Scharzfeld Cave, kept in the Royal Library, Hannover."

et saepe nitidi, et non raro portionibus maxillorum inserti; aliqui tantae magnitudinis, ut ad nota nobis animalia referri non possint.[81] Sub terra nigra flavus est limus, sed ossium expers, et subtus saxum, quod inquirentium curiositatem finivit. In hoc loco novissimo ipse vidi materiam alumini plumoso similem, quam incolae Stenomargam appellant, cui par quiddam in minerarum quarundam ustionibus attoli dictum est; facillime in pulverem abit fricando. Caeterum vix qinquaginta[82] annos esse audio, quod antrum Scharzfeldense detectum, aut certe celebratum est. Ex eo ossa ac dentes tota Germania in usum medicum circumferuntur; sed dum quisque pro arbitrio fodit, jam fere exhaustum intelligo angusto in spatio materiam curiositatis.

[XXXVII]

A Scharzfelda porro in Hercynias nostras valles ingressi ad Luderi montem venimus, quod oppidum est fodinarum haud expers. Fuit olim sedes Comitibus propriis, quos nostra Historia non silebit. Elbingerodae pernoctavimus. Postero die Brunlaegam itum, ubi ferri minera eliquatur, atque inde ad locum Rubeland. Ibi est Bumanni specus, quam nil solis egentes vesperi ingressi sumus. Prope Rubeland rudera spectantur veteris arcis Berckefeld, quae viciniam olim infertam[83] habebat. Sunt et ferrariae ibi, omnes Guelfebytani[84] juris. Post quas in edito monte antrum est Bumanni dictum, a primo exploratore, ferri venas quaerente, quem extincta lampade post aliquot dierum errorem egressum non diu supervixisse aiunt; etsi alii alia narrent. Ubi montem ascenderis primum subitur fornix naturalis.[85] Inde ad sinistram nonnihil descendenti aditus antri occurit, arctior illis, quibus fodinarum cuniculi exeunt. Inde ex angusto non sine difficultate eluctandum est in locum amplum, sed inaequalem, saxo nonnihil ab aquarum sedimentis incrustato stratum. Totam specum

---

81. In Hannover Ms XXIII, 23b, pl. 11 appears here.
82. B: quinquaginta.
83. B: infestam.
84. Guelferbytani.
85. In Hannover Ms XXIII, 23b, pl. 1 appears here.

indeed broken in pieces or scattered about, but you can still easily distinguish the body parts. There are many kinds of teeth of various colors; they are often shiny and, not infrequently, are still inserted into pieces of jawbone. Some are so large that they cannot be ascribed to any animal known to us. Beneath the black soil is a yellow mud, but it is without bones; and beneath that is rock, which puts an end to the curiosity of seekers. In this last place, I saw a material similar to plume alum, which the local inhabitants call *Steinmark*.[91] It is said that one gets something similar by burning certain minerals. When one rubs it, it easily turns to powder. Otherwise, I hear that it is hardly fifty years since Scharzfeld Cave was discovered, or at least much visited. Bones and teeth from it are sent throughout all of Germany for medical use. But since everyone digs as he pleases, I believe that this limited space is today already exhausted of interesting things.

[XXXVII. *The Baumann Cave and its contents*]

Moving on from Scharzfeld, we entered our Harz valleys and reached Lauterberg, a city not without mines. It was once the seat of certain counts, whom our history will not omit. We spent the night in Elbingerode. The following day we went to Braunlage, where iron ore is smelted, and from there to the village of Rübeland, where Baumann Cave is located. We entered the cave in the evening, since we did not need the sun. Near Rübeland, we saw the ruins of the old Berkefeld Castle, which once menaced the area. There are also iron mines there, and all of them are subject to the jurisdiction of the Guelphs. Behind these, on a high mountain, is the Baumann Cave, named after its first explorer. It is said that his lamp went out while he was looking for veins of iron ore, and he spent several days in the cave, wandering around, before emerging again; it is said that he did not survive much longer. Others tell it differently. When you ascend the mountain, you first encounter a natural vault. Descending a little to the left from there, one comes upon the entrance to the cave, which is somewhat narrower than a mine tunnel. From there, forcing oneself not without difficulty through the narrow passage, one comes into a wide but uneven cave, with walls composed of rocks that have been coated by sediments from the water. The whole cave is divided into five cavities, which

91. A kind of feather alum. See glossary.

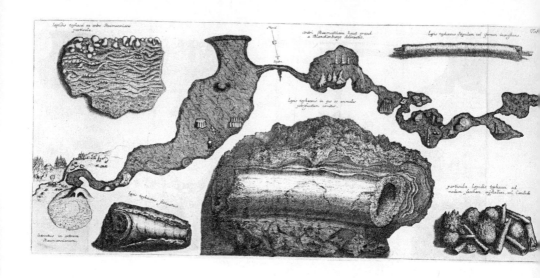

FIGURE 14. *The Baumann Cave*

Cross section of the Baumann Cave, located near the towns of Elbingerode and Blankenburg in the Harz. The original caption (*top center*) reads, "Sketch of the Baumann Cave, not far from Blankenburg." The plate includes depictions of fossil objects discovered in the cave. In his firsthand account of his visit to the Baumann Cave (*Protogaea*, §XXXVII), Leibniz noted the different details and curiosities he discovered along the way. The cross section that appears here—most unusual in its time—carefully notes the cave's orientation and displays various features: the entrance (*introitus in antrum Baumannianum*), the nearby ruins of the Berkefeld Castle, the successive chambers, several ladders in the narrow passages, and curious mineral formations, such as stalactites and stalagmites, in precise locations. The plate also contains several details that are mentioned in the text, like a fragment of tuff stone (*lapis tophaceus*) from the floor of the cave (*top left*), petrified straw encrusted with salt crystals (*top right*), and a piece of petrified bone enclosed in stone (*bottom*).

The engraving (plate 1 in Scheidt's 1749 edition) is unsigned. However, given the way it illustrates in detail the text of *Protogaea*, the plate was most likely produced according to Leibniz's own specifications.

in quinque cavitates, velut totidem antra, sed angustiis communicantia partiuntur. Inde transmissuro in cavitatem sequentem, quae amplissima est, rupi acutae inequitandum erat, cui ex eo nomen Caballi. Sed nunc iter commodius redditum est. A rupe descenditur per scalam, atque inde ulterius penetratur, ibi aliquot ex saxo columnas monstrant tanquam stillicidio aquarum formatas, et in una monachum, in altera Mosen bicornem videre se putant. Non procul a Mose, et alibi passim, occurrunt ossa belluarum et silices fluviatiles uno velut caemento involuti. Qui vero in his antris monstrantur naturae lusus imaginationis auxilio egent: Nam ascensionem Christi saxo expressam ostendunt, et baptisterium et furni similitudinem. Organon quoque musicum, et sylvam, et nescio quae alia. Prae ceteris elegans est tabulae species, in qua variae concretae figurae ac velut flores; hisque interspersi lapilli jacent incrustati; ut amygdali nucleos aut coriandri semina saccaro obducta putes. Sunt et stiriae anserini calami crassitie, variae longitudinis, aliquando tripedalis, quales si frangas, radii apparent candidi, crystallini a columnulae ambitu ad axem tendentes, quod etiam in belemnitis observes. Stiriae quaedam etiam ad calami modum cavae visuntur. Adsunt et columnae ingentes, in summo liberae, quae percussae ingentem sonum et velut campanarum aemulum edunt. Caeterum ut saxi naturam contemplarer accuratius, jussi frusta quaedam abrumpi, quae domi per otium examinarem. Ea considerans reperi, Spathi esse genus non dissimile nostro fodinarum, in quo et cavernulae, ingemmatura quadam circumquaque obductae, quemadmodum jam supra notatum est: Illud vero memoratu dignum visum est, saxum saxo inclusum, quod manifeste terminabatur crusta tenui obscure flavescente, qualis a recenti aquae illapsu lapidi inducitur, cui deinde circumdatum erat novo contextu aliud saxum, plane geminum priori. Ut appareant velut periodi, quales in arboribus annos definiunt, (Plinius pectines vocat), interquiescente scilicet natura, et post per novam illuviem opus resumente. Sed maxime me delectavit uni frusto, quod me vidente volenteque in antro abruptum erat, inclusa pars ossis, textura, superficieei folio et colore, denique et gustu prorsus suo, ut ab animali fuisse nemo spectator dubitare possit. Spithamae habet longitudinem, et utrinque apertum est, ut oculus alterutri foramini admotus videat diem. Serius animadvertimus, quam ut in eodem loco reliquum ossis, et, si quid aliud belluae restare poterat, vestigare nobis liceret.

are like five separate caves, though connected by narrow passages. Moving on to the next cave, which is the most spacious, one had to straddle a pointed rock that is called the horse for that reason. Today, however, the access has been made easier. From that rock, one descends a ladder and then goes deeper inside, where they point out several columns of rock formed by slowly dripping water. In one column, they think they see a monk; in another, Moses with two horns. Not far from Moses, and in various other places, are the bones of large animals and river pebbles, enclosed as if by the same mortar. But the games of nature presented in these caves demand the support of the imagination. For they point out the Ascension of Christ stamped in the rock, a baptismal font, something that looks like an oven, an organ, a forest, and who knows what else. Especially fine is a tablet upon which you can see various solid figures and imitations of flowers, and strewn among them little encrusted stones that you might take for almonds or sugarcoated coriander seeds. There are also stiria as thick as a goose feather quill and of various lengths, sometimes three feet long. If you break them, there appear white crystal spokes, radiating from the perimeter to the axis of the little column, as you might observe in belemnites.[92] Hollow stiria, like reeds, are also found. There are also huge, free-standing columns that produce a loud sound, like a bell, when they are struck. To consider the nature of the rock more carefully, I ordered that some pieces be broken off to examine with leisure at home. Upon examination, I found that they are made of a kind of spar not unlike that from our mines, in which there are also small cavities covered all around by sparkling stones, as noted above. But I found it especially noteworthy that there is a rock enclosed in another rock, which is clearly bounded by a thin, dark yellow crust that is even now still forming as water drips onto the stone. This crust is then covered by a new coat of rock, identical to the first. Something like the rings that designate years in trees (Pliny calls them *pectines*) thus appears, since nature rested intermittently and then resumed her work again with a new inundation.[93] But what pleased me most was a piece I found and then had broken off. For there is a piece of bone enclosed in it, and, on the basis of its texture, its surface, its color, and finally its taste, no observer can doubt that it came from an animal. It is one span long and open on both ends, so an eye at either end sees the light. I examined

92. The confusion between belemnites and stalactites was frequent in early modern literature.

93. Cf. Pliny the Elder, *Natural History*, 16.73–185.

XXXVIII

Res terram inter et mare, lapidem inter et lacrymam arboris ambigua Succinum est. Analysis chemica minerali regno favet. Et oleum inde petroleo cognatum paratur. Contra reperta intus folia, et muscus, et insecta, pro arboreo ortu pugnare creduntur: Sunt autem non nisi residua lineamenta, et velut umbrae rei inclusae, corpus ipsum dudum consumtum aperientibus nusquam occurrit. Hevelius, qui Borussiae suae velut haereditarium asseruit Astronomiae decus, caetera quoque naturae consultus, ad bitumen et gagatae similem naturam inclinabat. Vidi liquidius[86] adhuc, et sigilli capax. Libavius Medicus uno in frusto habuit et hanc ambram (nam et sic vocant) et adnatam illi alteram illam odore pretiosam, quae vulgo grisea dicitur, hic nigra erat. Nuper quoque ambram odoriferam in Borussia erutam intelligo. Verum ita non finita, sed translata quaestio est, cum hujus quoque genus ignoretur. Ad thermas Bellilucanas Galliae, fragmenta succini lapidibus agnati reperit descriptor. Et Goebelius melanteriam, aliaque subterranea succino adhaerentia vidit. Illud certum est, pene omne, quod recens advenit, colligiturque, ut in Borussiae potissimum litoribus, quanquam et Pommeraniae,[87] Frisiaeque non ignoretur, maris esse ejectamentum; itaque valde veresimile est, ubi nunc succinum effoditur obrutum arenis, olim mare vicinum fuisse. Repertum est autem non ita pridem in insula Albis, cui a Billa nomen, e regione Hamburgi a fossore cellam sub terra parante; et longius adhuc a mari ante aliquot annos ingens massa succini eruta est in villa Praefecturae Blumenau, non longe ab Hannovera, tempore Ioh. Friederici Ducis.[88]

86. B: Vidi et ego liquidius.

87. B: et Cimbriae, Danorum et Pomeraniae.

88. B: Circa Gartoviam, illustriss. Bernstorfii, Ducis Cellensis Status Ministri, oppidum, frequentius succinum in paludosa ibi regione fodientibus occcurit, et varia inde parata, ac inter caetera vasculum satis amplum ipse ego saepicule admiratus sum.

it afterward, so that I could no longer determine whether more bones or other animal remains were lying in the same place.

[XXXVIII. *On the nature of amber, especially the kind found in our region*]

Existing somewhere between land and sea, between stone and tree sap, amber is a puzzling thing. Chemical analysis favors the mineral kingdom. Also, oil is extracted from it that is similar to petroleum. Some believe, on the contrary, that the leaves, moss, and insects found in it argue that it arose from trees. But these are merely residual outlines, like shadows of the things enclosed; for the body is long since consumed, and one never finds it upon opening. Hevelius,[94] who staked a sort of hereditary claim to astronomical glory in his Prussia and was also a fine judge of nature in other respects, inclined to the view that the nature of amber was similar to jet and pitch. I have seen softer amber, into which one could stamp a seal. The physician Libavius[95] possessed in a single piece both this *ambra* (for some call it this) and, grown together with it, a precious-scented variety that one commonly calls "gray *ambra*."[96] In this case it was black. I hear that aromatic amber has recently been excavated in Prussia as well. The question is not, however, answered in this way, but only postponed, since the nature of this *ambra,* too, is unknown. Near the hot springs of Belliluc in France, amber that had grown together with the stones was found and described. Goebelius,[97] too, saw iron vitriol and other things sticking to the amber. Still, it is certain that almost all the amber found today, which is mostly collected in Prussia along the coast—though it is not unknown in Pomerania and Frisia—has been ejected by the sea. Therefore, it is very likely that the sea was once in the same area where amber is now dug out of the sand. Not long ago it was even found near

94. Johannes Hevelius (1611–1689).

95. Andreas Libavius (1550–1616).

96. The question of the nature of amber was not solved during the early eighteenth century, but Leibniz here notes the distinction between "succin" (i.e., yellow amber), which is a fossil resin, and "gray amber," which is produced by whales, commonly known as ambergris, and used to make perfume.

97. Johann Goebel (1683–1745), a Helmstädt jurist who edited the works of Conring.

XXXIX

Sed caetera ingentium naturae mutationum vestigia nonnihil tangamus habitatoribus fortasse antiquiora.[89] Aegyptum Nilo, Arelatensem agrum Rhodano deberi, Aristoteles et Peireskius credunt, Nannius Bataviam munus esse Boreae Rhenique. Certe flumina materiam advehentia spoliant superiores terras, Frisiique quotidie nostris detrimentis ditantur. Nec jam dico de insulis natis, qualem sub Leone Iconomacho supra memoravimus horribili terrae motu incendioque erupisse; nec de fretis a mari effractis, qualia Gaditanum et Siculum jam Veteribus judicantur; nec de montibus subversis, quemadmodum factum est in ditione Bernensi, et in Villaci Alpibus, et patrum memoria in Rheticis, cum Plursii oppidum opprimeretur, et in Firmano territorio, cum mons a cryptis dictus anno 1670. corruisset. Quanquam et apud nos in Blankenburgii tractus montibus, et alibi passim, manifesta sunt vestigia ruinarum, horrentia aspectu, quorum aliqua fluminibus imputes; praesertim in hoc loco Rostrap, ubi in procurrente scopulo ungulae, si Diis placet, notam ostendunt impressam rupi, unde in equo regis filiam cum amatore supra Bodae fluminis terribiles cataractas in oppositum montem transsiliisse poetantur. Passim etiam lacus monstrantur a ruinis et terrae motibus nati. Ut de Asphaltite Sodomae, et lacu Pilati, aliisque nil dicam; Steinhudensem in nostro tractu inter Leinam et Visurgim, vicini subsidente terra emersisse putant, quod illis facilius credo, quam quod urbem illic oppressam jacere, et fluctuum ejectamentis proditam, quasi a majoribus acceptum posteris tradunt.

89. B: antiquiora: Non illis tamen immorabimur, quae in nostris oris expressa non habentur.

Hamburg, on an Elbe River island called Billa, while an underground cellar was being dug up. And some years ago, in the time of Duke Johann Friedrich, an enormous mass of amber was dug up farther from the sea, in a village in the district of Blumenau, not far from Hannover.

[XXXIX. *Changes wrought by rivers and the vestiges of upheavals in our region*]

I would also like to touch on the other vestiges of nature's great changes, which are probably more ancient than her inhabitants. Aristotle and Peiresc[98] believe that Egypt arose out of the Nile and the region of Arles from the Rhône; Nannius[99] believes that Holland is a gift of the north wind and the Rhine. Rivers, which carry material with them, certainly rob higher terrain, so that the Frisians are enriched every day at our expense. I will not speak here about newly born islands like the one mentioned above, which burst forth through a horrible quaking and burning of the earth in the time of Leo the Iconoclast,[100] or about the straits ripped open by the sea, as the ancients concluded that Gibraltar and Messina had been. I will also not speak about mountain collapses, like what happened in the canton of Bern, in the Alps of Villach, and in the Rhaetian Alps, where, as our fathers remember, the town of Plüos was smothered, and also in the territory of Firmano, where the so-called Mountain of the Caverns[101] collapsed in 1670. Indeed, there are clear traces of destruction near us, horrible to behold, in the mountains of the Blankenburg region, and some of these can be attributed to rivers. This is above all the case at the Roßtrappe, where, if the gods please, they reveal a hoof printed in the rock of a protruding ledge.[102] Legend has it that a king's daughter, on horseback with her lover, leapt over the terrible waterfalls of Bode from

98. Nicolas Claude Fabri de Peiresc (1580–1637), collector, naturalist, and *conseiller* in the *parlement* of Provence. Peiresc corresponded with most of the *literati* of his time. Leibniz here refers to Gassendi's effusive tribute to him (Gassendi 1641).

99. Pieter Nanninck (1496–1557), humanist and professor at Leuven.

100. See n. 52.

101. The "Grottenberg."

102. The Roßtrappe and the Witch Dance Ground, located southwest of Quedlinburg in the Harz Mountains, remain popular tourist destinations to this day.

[XL]

Visurgim mutasse cursum in Mindensi tractu, atque olim sese infudisse paludibus a mari illuc usque porrectis, et ab Oceano aditum admittentibus, anchoramque etiam magnae navis ibi repertam incolae tradunt, sed rupto monte fluvium dextrorsum postea iter fecisse. Quod et Chronica quaedam Mindensia confirmant, quorum tamen autoritati in remotissimis parum tribuerim, nisi praesenti aspectu firmentur. Illud ne nunc quidem insolitum est, irrumpere Oceanum, aut repelli, aquasque et terras invicem permutari. Morinorum litus mari olim immersum fuisse, et ubi S. Audomari fanum est, Oceani portum extitisse Ortelius et Chifletius scribunt. Nec jam de Nordstrandiae inundatione in Holsatia, aut Belgica irruptione saeculi superioris dicemus, aliisque antiquioribus, cum mare visum est repetere jus suum, ne passim obvia inculcemus.[90]

90. B: Verum ut magis obvia inculcemus, lacus Steinhudensis in nostro tractu inter Leinam et Visurgim prodit, fere ad hanc usque urbem olim paludes ab Oceano irrugos pertigisse.

that place to the mountain on the other side. In some places, lakes that arose through earthquakes or collapses are also shown. I will not speak about Sodom's Lake of Asphalt,[103] Pilate's Lake, and others. One supposes that Steinhude Lake, in our region between the Leine and Weser rivers, emerged as the surrounding earth sank. I find this easier to believe than the tale, passed from ancestors to their descendants, that a sunken city lies there, as witnessed by debris from the waves.

[XL. *The struggle between sea and land*]

Local inhabitants report that the Weser River near Minden has changed its course; that it once flowed into marshes, which stretched to the sea, offering access to the ocean; and that the anchor of a large ship was found there. After a rockfall, however, the river changed its course to the right. This is also what certain *Minden Chronicles* confirm, which I nevertheless would consider of too little authority for such remote things, if the present appearance of the place did not support them. These days it is not uncommon for the ocean to intrude or be driven back, so that sea and land replace each other in turn. Chifletius and Ortelius write that the sea once covered Picardy and that there was a harbor where St. Omer now stands.[104] So as not to dwell on the obvious, we do not wish to speak here about the inundation of the island Nordstrand in Holstein,[105] about the sea's irruption into Belgium during the last century, or about other earlier times when the sea seemed to insist on some ancient right.

103. The Dead Sea.

104. In his *Portus Iccius Julii Caesaris demonstratur* (1626) Jean Jacques Chifflet (1588–1660) identified Portus Iccius, where Caesar once embarked for Britain, as St. Omer, today a town in northern France. Abraham Oertel (1527–1598), whose atlases enjoyed wide circulation during the sixteenth and early seventeenth centuries, made the same point. Both Oertel and Chifflet argued that Caesar embarked not from Calais, as is commonly assumed, but from St. Omer.

105. The Frisian island of Nordstrand, located off the coast of Schleswig-Holstein near the town of Husum, was inundated by the great floods of 1634, in which some eight thousand people died.

Magis operae pretium est in rem nostram, easdem apud Italos mutationes recognoscere in Estensium ditione, quam Brunsvicensium ducum
majores tenuere, et vicino Longobardiae et Venetiarum tractu. Et satis
quidem verisimile est, et Adriatici litoris Venetias (quae regionem olim
significabant, non urbem), et Aremorici a Caesare memorati (ubi hodie
Vannes) nomine suo communem originem ex paludibus atque illo genere
terrarum fateri, quod hodie Saxones, Belgae, Angli, Dani, vocant *Veen,
Fenne,* passimque siccatum in pascua abiit, et foenum praebet, interdum
et Turfam. Quantam autem mutationem tempus attulerit, fidem facis op-
positus historiae praesens vultus rerum. Constat magnam Adriatici litoris partem olim mari tectam aut paludibus inviam fuisse, et qui Aquilegia
Bononiam tendebant, magno ad dextram flexu olim usos apparet: Initio
scilicet Padus, et Athesis, caeteraeque illic Alpium et Appenini exoner-
ationes nondum ubique satis certis alveis coercebantur; donec Marcus
Scaurus manum admovit. Ex eo tempore certatum est cum fluminibus
optimique agri extorti, qui passim Italis policinia appellantur a paludibus: ut Rhodiginia, illa ad Athesin antiqua Estensium possessio, ubi et
sepulcra Majorum in Vangadigia habuere. Nunc ars eo progressa est, ut
alicubi videas terram mari, prata flumine depressiora, et aquam velut in
aere suspensam, cujus exundationem longissimus ager[91] coercet. Tantum
orbis facies mortalium studio mutata est, ut magnam habitationis suae
partem genus humanum credam ipsi sibi debere; tametsi et castores,
industrium animal, arte quadam sua, aggeres et stagna parare constet.
Apud Patavinos cum monasterii S. Helenae fundamenta locarentur, an-
coram quandam repertam Pignoria testatur, et in aliis urbis locis navium
malos. Ut credibile sit, usque ad Euganeos colles, quibus arx Atestis, et
mons Silicis incumbunt, maris aestuaria aut stagnantia in exitu flumina
pervenisse. Ravennam ita describit Strabo, ut quis hodie posset urbem
Venetiarum,[92] quarum perviam mari lembisque talis adhuc Cassiodoro
fuit pro Theodorico Rege scribenti. Inclinante in Occidente imperio erat
ibi primaria statio Romanae classis, et Exarchi sedes, quod commercio
navigandi Graeciam Italiae connecteret. Nunc mare dudum oppletis aes-

91. B: agger.

92. B: urbem Venetiarum, scilicet perviam mari lembisque. Immo talis adhuc Cassiodoro fuit pro Theodorico Rege scribenti.

It is more valuable for our subject to recognize similar changes in Italy, in the domain of Este,[106] a territory near Venice and Lombardy once possessed by ancestors of the most serene dukes of Brunswick. It is very likely that the Venetians along the Adriatic coast—for the term Venetias once denoted a region, not a city—and the Aremorici[107] mentioned by Caesar (where Vannes is today) indicate through their names a common origin in marshes and the sort of earth that Saxons, Belgians, English, and Danes call *Veen* or *Fenn* which, when drained in spots, becomes pasture, yielding hay and sometimes peat. But you can be sure how much change time has wrought by comparing the present face of things to what history describes. Certainly the sea once covered much of the Adriatic coast, or marshes made it impassable. And whoever wanted to travel from Aquileia[108] to Bologna clearly had to make a big detour to the right. For initially the Po, the Adige, and the other rivers that drain the Alps and the Apennines were not completely contained in their beds, but then Marcus Scaurus[109] applied his hand to the matter. Since that time one has battled with the rivers, wresting from them extremely fertile lands, which in Italy one generally calls *polocinia,* from the Latin *paludes* (marshes); this is true in Rovigo, that ancient possession of the House of Este along the Adige, where the tombs of their ancestors lie at Vangadizza.[110] These days art has advanced so far that, in some places, you see land lower than the sea, meadows below a river, and water that seems suspended in the air,

106. The "region of Este" had its center in Ferrara, over which the family of Este ruled between the thirteenth and sixteenth centuries. Ferrara lies some fifty miles southwest of Venice along the river Po.

107. "Aremorici" is a Gaulish term meaning "those who live beside the sea." Armorica is an older name for Brittany.

108. Aquileia, now a village near the Adriatic coast northwest of Trieste, was founded as a Roman colony in 181 BCE and later became an important patriarchate of the Roman Catholic Church.

109. In 109 BCE, Marcus Aemilius Scaurus (162–89 BCE) directed construction of the Via Aemilia, which passed through Pisa to Dertona (Tortona), a town situated just west of the northern Apennines.

110. The Polesine, one of the richest agricultural regions of Italy, was the southernmost of Venetian districts, situated between the final stretch of the Po and Adige rivers near the Adriatic. Rovigo was capital city of the Polesine, and Vangadizza was a famous abbey there.

tuariis recessit; eandemque temporis injuriam et Veneti timent; quos sunt, qui remedii spe, averso et fracto fluminum impetu, malum auxisse arbitrentur.

[XLII]

Ingentem velut lacum terra obrutum, immo urbe et agro, velut fornice, opertum, insigni naturae omnia vertentis miraculo, sub se sentit Mutina, Estensium hodie Principum sedes:[93] nec aliquid tota Longobardia temere se offerat dignius describi. Habet scilicet Mutina, quod nescio, an alius orbis locus, ut tota urbe, et extra quoque in vicino aliquo usque agro, ubilibet liceat fontem facere, vivum, salientem, perennem, et, ut verbo dicam, rivulum artificialem, cui nec detrahat aestas, nec addat hyems. Neque aliud postulant Aquileges, quam putei fodiendi locum, qui septuaginta pedum altitudine deprimitur. Et initio quidem ad decem pedes rudera occurrunt veteris urbis, et tesselatum opus stratarum olim platearum, aliaque subinde antiquitatis vestigia eruuntur. Tantum urbis suis ruinis et terris advectis crevit, inde simplex terra quattuor aut quinque pedum; tum iterum rudera in duodecim pedes, quasi urbe plus semel subversa. Inde *creta*, ut vocant Itali, id est argilla tenax pedum XXIV. Tum

93. B: hodie Serenissimorum Estensium sedes.

restrained from overflowing by a very long dam. The face of the globe has been transformed through the efforts of so many people that I believe humans owe a great part of the land they inhabit to themselves, although beavers, those industrious animals, also construct dams and ponds according to their own special art. Pignoria[111] reports that, while laying the foundation for the monastery of St. Helen in Padua, they discovered an anchor, and that they found ship masts in other parts of the city. It is likely, therefore, that estuaries of the sea or lakes formed by rising streams have reached up to the Euganean Hills,[112] upon which the Castle of Este and Mt. Silicis rest. Strabo described Ravenna just as someone would describe the city of Venice today; and Cassiodorus, who wrote for King Theodoric, described Ravenna as accessible to the sea and to small ships.[113] During the decline of the Western Empire, the city served as first station of the Roman fleet, and the seat of the exarch, because it joined Italy to Greece through maritime trade. Now the sea has receded, with the estuaries long since filled. The Venetians fear the same ravage of time, and some claim that those who, in search of a remedy, diverted and broke the assault of the waters, aggravated the damage.

[XLII. *The marvelous fountains of Modena*]

Because of a striking wonder of all-changing nature, one discerns under Modena, present seat of the princes of Este, a vast lake hidden in the earth, covered by city and field as if by a vault. Nothing in all of Lombardy is more worthy of description. I know of no place in the world like Modena: wherever one proposes to sink a well, throughout the whole city or in the neighboring fields just outside it, one has a living, springing, continual and, in a word, ready-made stream that neither decreases in the summer nor increases in the winter. The well diggers ask only about the location of the well shaft, which they sink to a depth of seventy feet. Initially, at

111. Lorenzo Pignoria (1571–1631), known for the collections of books and curiosities, especially Egyptian, that he assembled in Padua.

112. The Euganean Hills, a group of volcanic outcroppings of relatively recent origin, lie about twenty kilometers southwest of Padua.

113. Flavius Magnus Aurelius Cassiodorus Senator (ca. 490–583 AD), Roman writer and monk who served as councillor to Theodoric the Great, king of the Ostrogoths.

alia terreni species, cui valli nomen fecere, quam olim detectam fuisse radices arundinum, et fracidi stipites, et rami truncique, et folia arborum, et interspersa passim conchylia fatentur. Congesta est haec terra foliorum seu pectinum instar, pluresve limbi sibi superstrati noscuntur, manifesto inter eos terrae purae et fracidae materiae discrimine. Sub hoc vallo, cujus altitudo rursus est XXIV. pedum, terra est argillosa iterum, ad quatuor pedes, multo priore tenacior. Postremo mixta arenae glarea occurrit, murmurque auditur ingens, et fremitus, velut labentium aquarum. Tum vero fossor ita se parat, ut terebrae insidens suae, attolli promte possit. Nec mora, dum pergit ferrum, ecce aqua prorumpit ex solo, lentior primum, et mista arenae, mox tanta vi, ut vix retracto homine, insecuta assurgat ad summum putei labrum, indeque erumpens continuum faciat profluentem, tecto sub terra cameratoque lapidibus rivulo, usque ad communem alveum omnium fontium urbis, quem canalem magnum appellant, qui ad Panarem flumen ducit. Haec coram accepi vidique, quae Bernardus Ramazzinus, insignis doctrinae medicus apud Mutinenses, in justum opusculum eleganti mechanicae pariter et naturalis scientiae specimine parat. Additur, terebram, cum aquas attingat, semper vi actam versus Ferrariam declinare, quasi illuc labentibus sub terra aquis in fluminis modum. Sed hoc minus compertum haberi, aut potius falsum esse, Ramazzinus notavit. Hyeme non foditur, ob molestum putei calorem, quem suasi, ut imposterum thermometro explorent, ne forte pro Antiperistasi suffocantis in loco non pervio aeras natura imponat. Aqua purissima est, et prae caeteris ex fonte Abyssi, quem vocant, hauritur, non inferior Nuceriana. Puteus semel factus, aeternum est beneficium loci, tanta perennitate, ut nec labore hominum exhauriri potuisse compertum sit; quanquam saepe tentatum constat, impurius aliquando fluente aqua, dum forte imum putei os male foratum, aut postea oppletum est. Ubi non aliud remedium habetur aut requiritur, quam ut demissa per medias aquas terebra repetat officium.

a depth of ten feet, they encounter the debris of the old city, unearthing the paving stones of forgotten avenues and other vestiges of antiquity. So far did the city rise on hauled earth and its own ruins. Next come four or five feet of simple earth; then debris again twelve feet lower, as if the city had been destroyed more than once. After that comes what the Italians call *creta*, that is, a tenacious clay twenty-four feet thick; then another kind of soil, which has been called *vallus*. Reed roots, rotten trunks, twigs and stems, tree leaves, and the shells scattered among them declare that this *vallus* was once exposed. This earth has piled up like leaves, or the teeth of a comb, and one discerns many strips lying upon each other, so that it is easy to distinguish between pure earth and rotten matter. Beneath this *vallus*, whose depth measures another twenty-four feet, come four more feet of argillaceous earth more tenacious than before. Finally, one encounters sand mixed with gravel and hears a mighty growl and roar, just like rushing waters. Then the digger prepares himself suitably, so that, sitting on his auger, he can be lifted out quickly. Soon, as the drill continues, see the water break out of the ground, slowly and mixed with sand at first, but soon with such force that the man is scarcely drawn back before the water surges up behind him to the highest lip of the well! And springing forth from there, it generates a continual flow through a subterranean stream vaulted by stones and into the general conduit for all runoff from the city's wells, called the Great Canal, which empties into the river Panaro. I observed this and saw for myself what that exceptionally learned physician in Modena, Bernardino Ramazzini,[114] describes in a fine little work, which provides a model both for mechanics and natural philosophy. In addition, it is noted that the auger always bends toward Ferrara when it touches the water, driven by force, as if the subterranean waters flowed like rivers. Ramazzini observed, however, that this had hardly been proven, or rather that it was mistaken. One does not dig during the winter because of the unpleasant heat of the well, so I recommended using a thermometer to investigate whether nature might

114. Bernardino Ramazzini (1633–1714), often called the founder of occupational medicine, became chair of medicine in the reopened University of Modena in 1682. He taught there for eighteen years before leaving for Padua. Leibniz is probably referring here to Ramazzini's 1691 work *De fontium Mutinensium admiranda scaturigine*.

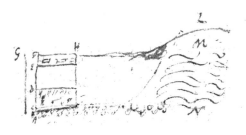

De terrae stratis non difficilis conjectura est, primae glareae ex altioribus locis infusam aliquando argillam, huic fortasse arundineta et palustrem materiam cum arboribus super increvisse, diuturno temporis intervallo; novo deinde maximo impetu immensam argillae vim advectam.[94] Huic inaedificatam veterem urbem subversam irruptione barbarorum atque ab eo tempore terreni molem pluviis torrentibusque crevisse. Scimus Romae nunc in Pantheon Agrippae descendi, in quod olim gradibus aliquot ascendebatur, remota non ita pridem terra, ut ima columnarum detegerentur. Aquileiae audio plus semel alternantes cum terra ruinas deprehendi, quasi facie urbis toties mutata. Sed aquae salientis, quod diximus, Mutinense miraculum majus negotium facessit. Qui fluminis labentis impetu expelli suspicantur, non satis attendunt, quanta vi opus sit ad fontes continuos in altitudinem septuaginta pedum propellendos. Multo majore scilicet necessaria velocitate, quam lapis acquireret lapsus ex altitudine eadem. Et, si tanta esset cursus pernicitas, non utique sursum verteret directionem. Praeterea glarea arenisque manifeste impeditur: murmur certe non statim a cursu aquae est, sed qualincunque fluctuatione circa locum pro-

94. B: vim advectam, cui inaedificata vetus urbs irruptione Barborum subversa. Ita enim ab eo tempore terreni molem pluviis torrentibusque crevisse facile concipimus.

deceive through antiperistasis[115] in poorly ventilated, suffocating places. The water is most pure, especially what is drawn out of the so-called Fountain of the Abyss, which rivals the water of Nocera. Once made, a well provides lasting benefit to a place, and of such permanence that human labor can certainly never exhaust it, though no doubt it has often been attempted, as when, occasionally, the water flowed impurely, because the lowest opening of the well shaft had been dug badly, or had become clogged afterward. Then one has or seeks no other remedy than that the borer is lowered through the waters to repeat his duty.

[XLIII. *How Modena's fountains are produced*]

It is not difficult to conjecture about these layers of earth. Clay from higher places once poured down onto the initial gravel; then perhaps trees covered over reeds and swampy matter during a long interval of time; and finally came the vast quantity of clay, conveyed by a strange and most violent impulse. Upon this foundation was built the old city, which the barbarian invasions destroyed; after that, one perceives how rains and floods heaped earth onto the ruins. We know that whereas in today's Rome one climbs down into the Pantheon of Agrippa, it was once necessary to ascend a few stairs, for the lowest of the columns was only recently excavated. I hear that one observes alternating layers of earth and ruins in Aquileia, as if the city had altogether changed its appearance as many times. But the leaping waters, which we called the miracle of Modena, present greater difficulty. Those who suspect that these waters are ejected by the power of falling water do not adequately consider how much force there must be to drive unbroken fountains to a height of seventy feet. Clearly, the requisite velocity[116] is much greater than what would be acquired by a stone that had fallen from the same height. And even if the swiftness of the current were that great, it would clearly not

115. Antiperistasis, a term from Aristotelian physics, denoted the process in which, given two contrasting qualities, the opposition of the one increased the power of the other. "Thus cold, say the school-philosophers, on many occasions exalts the degree of heat" (Chambers 1728, vol. 1, 111).

116. The distinction between speed, a scalar quantity, and velocity, which had direction, was crucial to Leibniz's physics and metaphysics. Cf. Garber 1995, 284–293.

fundum et cavum. Et quanquam exonerari aquam non dubitem subterraneis exitibus, lente tamen fieri credendum est, nullo alias hydrophylacio pro tanta latitudine ac celeritate suffecturo. Itaque non video, quid aliud restet, quam vicino monte tegi ingentem lacum, hunc antiquissimis temporibus ad locum urbis pervenisse, ubi glareae fundo incumbebat. Deinde super injektam materiam per partes ex ipso cacumine devectam, expleta demum cavitate, sed ita ut per intervalla lapillorum argillam tenacem velut fornice sustentantium communicatio aliqua cum lacu salva perstaret. Unde inaequali licet urbis solo, ad ejusdem tamen horizontis libellam ubique pervenit aqua, eandem scilicet, quae est lacus. Nam superiore urbis loco putei per se exonerantur in apertum, infra vero subterraneos exitus habent.

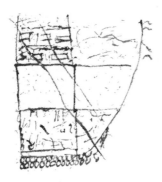

[XLIV]

Ab Estensi Longobardia revocamur in Brunsvicensem tractum non absimili ruina terrarum, cujus saepe deprehensa vestigia sunt, sed illustre imprimis documentum nuper vicinum Goettingae Rostorfium dedit, nobilium olim dynastarum sedes. Rem ex Parochi, non indocti viri, narratione huc referre operae pretium est. Incola loci puteum fieri curaverat. Primo humus consueta frugifera duodecim pedes alta perfossa est, deinde nigra ex putrefactis foliis, fibris, musco, multis referta conchis, spatio unius pedis: Sub qua margam invenit putearius trium fere pedum altitudine, perviam eanalibus scaturiginosis. Huc usque prima vice perductus est puteus: Sed cum nuper mense Aprili anni supra octogesimum sexti aqua defecisset, altius in profundo quaeritur scaturigo. Ibi Fossori iterum offertur nigricans et grave olens solum ex putrefactis foliis, stipula, gramine, radicumque fibris, multisque conchyliis octonum pedum densitate. Sub hoc demum ubi argillosus incipit et lubricus fundus in betulam offendit fere putrefactam densamque abietem, adhuc integram ex transverso jacentem cum radicibus suis, conisque nonnullis juxta repertis. Nec dubium est, vallem arboribus consitam cum nondum frequentia hominum excisa in his oris nemora essent, illuvie aquarum per intervalla

reverse its direction upward. Moreover, gravel and sand clearly hamper the flow. The rumbling certainly does not come from running water, but rather from undulation around a deep and hollow place. And though I doubt not that the water drains through subterranean passages, it is at least credible that it happens slowly, because otherwise, no water chamber would suffice to supply such a swift and broad current. Therefore, I see no option except that the neighboring mountain concealed a huge lake, which, lying on a bed of gravel, stretched to the site of the city in the most ancient times. After that, the material embedded above fell down from the peak of its own accord, piece by piece, so that it ultimately filled the hollow. But this happened in such a way that the spaces between the rocks, which supported the tenacious clay like a vault, retained some connection to the lake. And though the ground of the city be uneven, water everywhere reaches the same horizon level, which is that of the lake. For in the upper part of the city the wells drain themselves openly, but they have subterranean outlets below.

[XLIV. *The layers of earth in Rosdorf, near Göttingen*]

Este in Lombardy recalls an analogous ruin of earths, whose traces are often discovered, in the region around Brunswick; Rosdorf near Göttingen, ancient seat of a noble family, recently yielded up a particularly fine example. It is worth the effort to report this thing after the account of the pastor there, a not unlearned man. A local inhabitant had a well sunk. First one dug through the customary twelve feet of fertile soil; then came dark earth, one foot thick, and composed of rotten leaves, fibers, moss, and crammed with many shells. Under this the well digger found marl three feet thick, permeated by arteries of spring water. The well advanced this far the first time. But recently, in April of 1686, when the water had abandoned the surface, its source was sought at a greater depth. There the diggers again encountered a black stinking soil, eight feet thick, composed of rotten leaves, stalks, grass, root threads, and many shells. Under this, where argillaceous and slimy ground began, one hit upon an almost rotten birch and a solid fir, still whole and lying crosswise together with its roots, and with some fir cones next to it. There can be no doubt that the valley, when it was not yet populated and before the forests within these borders were chopped down, was once filled with trees. Repeated

redeunte, diversi generis stratis oppletam; cum et marga illic colles abundent, et fibrae radicum ac folia, manifesta spolia sint astantium sylvarum. Nunc locus ita editus est, ut ingens ex imbribus inundatio, quae sese velut ruptis nubibus quadriennio ante effuderat, magno in depressionibus damno dato, ad putei locum non pertigerit. Nec memoratu indignum est, in illo tractu nunc abietes desiderari; tantum natura loci mutavit. Conchae non ex bivalvium genere sunt, sed turbinatarum: nec dubitare fas est, veras esse, et vivo olim animali habitatas.

[XLV]

Passim alias occurrunt arbores obrutae, et fossile lignum; quale in Umbria repertum Franciscus Stellutus peculiari opera tractavit; nec limo tantum mersum, sed et saxo obvolutum scimus. Ex Chronico Montanorum Misniae constat, fagum cum ramis et foliis in saxo cinereo durissimo sub terra altitudine centum et octoginta ulnarum repertam, et cornu uri in profundissimis Thuringiae cavernis, quin et sudem in sepis usum paratam occurrisse. Accepi etiam die Februarii septimo, anni aerae Christianae millesimi sexcentesimi quinquagesimi sexti ad vallem Joachimmicam, oppidum Bohemiae fodinis celebratum, repertam in cuniculo a Barbara Brulla cognominato, ad profunditatem centum et quinquaginta orgyiarum petrificatam quercum cum radice ramisque. Fagus, unde cotes fiebant, ex ejusdem oppidi fodinis septuaginta orgyiarum profunditate Gesnero et Albino jam olim memorata est. Et nunc deliberandum relinquo, an in tam humili loco natam arborem, deinde vertenda in saxum materia oppressam, an ruptis terrae veteris fornicibus ex montibus cum massa ambiente in cavitates devolutam putemus? Benjamin Olitschius, mineralium egregius indagator, qui apud nostros rei metallicae consultus, mox a Batavis Gubernator fodinarum in Orientalem Indiam missus, praemature obiit, inter collectanea sua habere mihi significavit lapidem ab Augustoburgo Misniae, quem agnosceres ex alno factum. Et alium, quem *Schlamstein* vocant, quasi ex limo productum diceres, ubi in eodem frusto ita expressa erant duo folia, ut species discerneres, unumque quercus, alterum salicis faterere. Neque ille verorum ex suis arboribus lapsorum vestigia fuisse dubitabat. Hunc acceperat ab Ephippio vetere

watery deluges then filled it with different kinds of layers, for the hills abound in marl, while root fibers and leaves are the clear remnants of standing forests. Now, the place is so elevated that even the huge flood which came pouring out of cloudbursts four years ago and inflicted damage on the lower-lying places did not reach up to the site of the well. Nor is it unimportant to recognize that the area now has no fir trees. So much has the nature of the place changed. The shells are not of the two-shelled kind (bivalves) but rather of the conical variety (snail shells), and one cannot doubt that these are true shells that were once inhabited by living animals.

[XLV. *On buried trees and petrified wood*]

Elsewhere one finds fallen trees and petrified wood, which Francesco Stelluti[117] discovered in Umbria and described with singular effort, and which we perceive not only covered in mud but also enveloped in stone. It is well known from the *Meissen Berg-Chronica*[118] that a beech tree was found, complete with leaves and branches, one hundred and eighty ells[119] under the ground in very hard, ash-colored stone. A stake used for fence making was also encountered in the deepest Thuringian caverns, next to a bison horn. I also heard that a petrified oak was discovered, with its roots and branches, one hundred and fifty *Lachter*[120] down, in a tunnel called Barbara Brulla on 7 February 1656 in Joachimsthal, a famous Bohemian mining town. As Gessner and Albinus[121] once recounted, there was a beech tree, from which they made whetstones, seventy *Lachter* underground in the mines of this same town. And now I leave open for consideration

117. Stelluti (1577–1651), a member of the Accademia dei Lincei, worked with Frederico Cesi to collect and describe Umbrian fossils, especially fossilized wood. His *Trattato del legno fossile minerale* (1637) argued that petrified woods were minerals, and not the remnants of living plants.

118. Cf. Albinus 1590.

119. An ell was roughly a yard, but the measure varied widely from place to place; the English ell, for example, was considerably longer than the Flemish ell.

120. An old unit of measure used in the mines. One *Lachter* was roughly two yards (or two ells).

121. Konrad Gessner (1516–1565), Swiss naturalist. See Ogilvie 2006, 34–36, 236–240. Petrus Albinus, a.k.a. Peter von Weisse (1543–15989), professor in Wittenberg.

(*Altensattel*) loco Bohemiae, ubi etiam non procul Egra flumine integri arborum trunci in saxum versi deteguntur. De Ebeno fossili nostrate jam Cordus Agricolae retulerat, Hildesheimii intra terram aluminosam esse lignum in lapidem mutatum, idque in saxi commissuris reperiri. Idem habet in eadem terra aluminosa lignum quernum petrefactum. Addit e regione arcis Marieburgae collem esse plenum lapideis trabibus, quarum capita interdum eminent. Esse vero perlongas acervatim positas, inque medio earum terram colore nigram; de quibus tamen ego nil pronuntiare audeo, re nondum satis excussa. Agricola ad saxeum genus inclinat. Conringius eos, qui coram inspexere, lignis potius adscripsisse refert. Quod addit Agricola, ferro aut alio lapide percussas trabes, quemadmodum et Ostraciten ejusdem loci, cornu usti virus olere, id cui rem tribuam, non satis scio. Nam inest aliquando et mineralibus odor ex animali aut vegetabili regno. Et ne urinosum quiddam spirantes referam sulfureas aquas, locaque ubi sal excoquitur, scimus saxum esse violas olens prope Altenbergam, fodinis stanni celebrem, ac prope Silesiacam Herzbergam. Et fossoribus constat Coboltum, unde Bismuthum, Zafera et Arsenicum parantur (ex uno lapide omnia), quodam allii odore in ipsa vena prodi. Unde aliquando suspicatus sum, ex Knoblochio, quod Germanis allium est, corruptum Cobolti nomen.

whether we should suppose that the tree grew in such a low place before being buried by petrifying material, or, rather, whether it tumbled from the mountains into the depths, together with the surrounding matter, when the vaults of the old earth ruptured. Benjamin Olitsch[122] told me he had a stone from Augustusburg in Meissen in his collection that you could recognize as being made out of an alder tree; he was an excellent investigator of minerals, who served as a mining councillor for us before the Dutch sent him to the East Indies as director of mines, where he died prematurely. He had another stone they call *Schlammstein*—just as if, you might say, it had been formed out of mud; a piece of the stone had two leaves so clearly impressed on it that you could discern what kinds they were: the one an oak, the other a willow. Nor did he doubt that these were the traces of real leaves, which had fallen from these trees. He got the stone from Old Saddle (Altensattel),[123] a place in Bohemia not far from the Eger River, where one uncovers entire tree trunks turned to stone. Cordus once told Agricola that our fossilized ebony was wood that had turned to stone inside the aluminous earth of Hildesheim, and that one found it in the seams of the rock. Likewise, the same aluminous earth contains petrified oak wood. He adds that there is a hill full of stony timbers, whose ends sometimes jut out, near Marienburg Castle. They are in truth very long and piled in heaps, with black earth between them. I do not, however, venture to judge about this, for the thing has not yet been examined enough. Agricola was inclined to regard it as a kind of stone. Conring relates that those who have inspected the material in person tend to identify it as wood. Agricola adds that these timbers emit the smell of burnt horn when struck with iron or another stone, just like the ostracites in that place; how I might explain that, I don't really know.[124] For there is sometimes in minerals the smell of the animal or vegetable realm, not to speak of sulfuric waters, which give off a certain smell of urine, and places where salt is boiled; we also know that there is a rock that smells like violets near Herzberg in Silesia and Altenberg, a place known for its tin mines. And it is well known among miners that cobalt,

122. Olitsch was sent to Sumatra as a mining director (*Berghauptmann*). He died there, with almost all the miners who accompanied him, in 1682. Leibniz received word of his death in 1683. Cf. Leibniz [1683] 1923–, ser. 1, vol. 3, 563, 591.

123. Leibniz is probably referring here to Altsattel, or Stare Sedlo, a town in the mountainous regions of western Bohemia.

124. Cf. Agricola 1546, bk. 7.

Caeterum et sub Torfa, quae non procul Hannovera Cellaque effoditur
in urendi usum, veteres passim arborum trunci reperiuntur, et velut filamenta ligni. Torfa autem non terra est, sed materiae vegetabilis colluvies,
forte ex erica, musco, gramine, radicibus, arundinibusque terrae palu-
dosae postremo siccatis longissimo tempore concreta. Nec video, cur
sulphur et bitumen huic magis quam caeteris foci alimentis inesse necesse sit. Etsi aliquibus speciebus admista esse possint. Torfae artificialis
genus coriariis ex corticum quernorum reliquiis domi suae nascitur. Vix
apud nos hominis altitudinem excedit ustilis materiae stratum; excisam
renasci nondum compertum, etsi aquae advehant in vicinis locis jam natam. Maggenbergae in Misnia, ad nigrae aquae rivum, quarta orgyia venam ferri dedit, eaque perfracta apparuit ustilis cespes. Chauci, Bructeri,
Frisii, Cimbri, Belgae, Picardi Turfa utuntur, nec in Anglia deesse audio;
non magis quam sub ea subrutas arbores, quas ibi *Mosswood* vocant, quasi
lignum sub musco. Adeo non soli populorum terram nostram urimus,
quod Plinius exprobravit. Videntur usum homines ab incendiis didicisse,
quae aliquando late vagantur per hoc soli genus, diuque durant; quale
sicca aestate vix mensis spatio restinctum annales Bremenses memorant
anno seculi a Christo duodecimi septuagesimo octavo. Non omnis tamen
Torfae eadem natura: Apud Seelandos Belgas Darvia maris ejectamentum
est, cujus cineribus sal vescus elici potest, quod aliquamdiu prohibitum,
ne terra firmitatem coercendo mari necessariam fodiendo amitteret, Car-
olus V. pauperum precibus rursus indulsit. In Batavis vasti passim campi,
quae *Veenas* vocant, ubi remota crusta ad materiam Torfae subterraneam
pervenitur, limosa specie ob aquarum admistionem. Hanc haustam per
solum extendunt, et humorem pedibus, interjecto assere, exprimunt ad
justae consistentiae firmitatem. Inde in parallelepipeda seu lateres sec-
tam formant, soleque siccant et ventis, ad usum urendi. Qui terram pub-
licam Torfae commercio sibi vendicant, pro fructu paucorum annorum
perpetuum sibi onus quaerunt solvendi de agro inutili census. Longum

from which one prepares bismuth, arsenic, and zaffer[125] (all from one stone), betrays itself through a certain garlic smell in the ore vein. I once suspected, therefore, that the term cobalt is a corruption of *Knoblauch*, which is German for garlic.

[XLVI. *Peat and its origin*]

Moreover, peat is extracted for fuel not far from Hannover and Celle, and beneath it one sometimes finds old tree stumps and, as it were, woody threads. Still, the peat is not earth, but a hodgepodge of plant material— heather, moss, grass, roots, and reeds—that accidentally coalesced on swampy ground and then dried during an extremely long period of time. I do not see why there must be more sulfur and pitch in peat than in other fire-feeding materials, even though these materials can be mixed with various sorts of things. The tanners generate an artificial kind of peat in their workshops from scraps of oak bark. Near us, the layer of burnable material hardly exceeds the height of a person; it has not yet been observed that peat grows anew after being extracted, though the waters might convey it, already formed, to nearby places. At Maggenberg in Meissen, along the brook Schwarzwasser (black water), an iron vein was discovered at a depth of four *Lachter*, and, having broken through this, there appeared burnable turf. The Bremenites, Westphalians, Frisians, Holsteiners, Belgians, and Picards[126] make use of peat, and I hear that it is not lacking in England, where one also finds trees under the earth called "Mosswood," like wood under moss. So we are not the only people who burn our earth, a practice censured by Pliny.[127] It appears that the fires, which sometimes range widely through this kind of earth and last for a long time, taught people to use it. The Bremen chronicles of 1278 report such a fire, which was hardly extinguished in the space of a month during a dry summer. Still, not all peat is of the same nature; the sea spits out a peat called *Darvia* near Belgian Zeeland, and one can

125. A pigment (impure cobalt oxide) used to color porcelain and enamel.

126. Leibniz here uses the names of Roman tribes to connote peoples living in seventeenth-century Europe. Since many of these tribes, like the Cimbri, were nomadic, the choice of modern equivalents remains somewhat arbitrary. For a detailed breakdown, see Jean-Marie Barrande's notes: Leibniz 1993, 161 nn. 1–4.

127. Pliny, *Natural History*, bk. 7.

esset expectare,[95] dum Torfa renascatur orbe alio post Platonicam rerum revolutionem. Locum ejus interim aqua opplet paludosa, nec profutura, cum abduci non possit. Itaque qui negotium in se suspiciunt, ita calculos ponere debent, ut praesentanea utilitas non impensis tantum, sed et sorti sufficere possit, cujus usura annum canonem excedat. Turfa,[96] quae passim per Chaucos, et Cheruscos, et Bructeros, et Morinos uritur, seseque ad Somam fluvium extendit, plerumque in aperto est. Nec abhorreo a probabili conjectura inundationem esse foetum. Nempe semisiccato post aquarum illuviem solo, tenuis ericae rudimenta velut velatum increvere; mox nova inundatio, novique limi subtile sedimentum; rursusque in hoc ericae novae stamina ducta; donec post multas annorum vicissitudines, in praesentem crassitiem ustilis cespes augeretur; cessantibus tandem incrementis, ex quo flumina sibi viam magis magisque excavantia, aliquando et coercita humano labore, certiore jam alveo fluere coepere. Confirmant hypothesin plaggae, id est superior terrae crusta, a nostris Westfalisque ruricolis, de nudato sabulo ericetorum abscissa, partim ad agros steriles utcunque amendandos, partim in foci usum. Itaque ustilis est ericae superficies, ob terram plantulis herbescentibus interstinctam; Turfa[97] autem velut replicatione plaggarum accrevit.

95. B: nec forte hoc contingent, nisi in orbe alio post Platonicam rerum revolutionem.
96. B: Torfa.
97. B: Torfa.

make table salt from its ashes. This was forbidden for a time to keep the earth from losing the necessary firmness to restrain the sea. Because of the entreaties of the poor, Karl V allowed it again. In Holland there are empty fields all around, which they call *Veene*, where, after having removed the outer crust, one encounters a subterranean peat that is boggy because it is mixed with water. They draw it out, spread it on the ground, cover it with a board, and press out the water with their feet, until it becomes suitably firm. Then they carve it into parallelepipeds or tiles and dry it with the sun and wind for burning. Those who appropriate public land in order to trade in peat assume the burden of paying perpetual taxes on a useless field in return for a few years' profit. You would wait a long time before the peat was born again, in another cycle and after a Platonic upheaval of things.[128] In the interim, the empty space fills with swampy water, which is useless, because one cannot drain it. And those who undertake this business must therefore fix their calculations so that the immediate return not only covers expenditures, but can also furnish a principal whose interest will exceed the annual tax. Peat, which is burned all around Bremen, the Weser basin,[129] Westphalia, and Flanders,[130] extends to the Somme River and generally lies in the open. I do not reject the likely conjecture that it was the offspring of floods. Of course, after the flood of waters, sparse young heath grasses spread like a veil over the half-dry earth; soon came a new flood and another deposit of fine silt. New heath-grass fibers once again sprouted on this, until finally, after many years of such cycles, the burnable turf grew to its present thickness. At last these increases came to a stop, because, more and more, the rivers carved out their own road, sometimes even restrained by human toil, so that they began to flow in established beds. This hypothesis is confirmed by the *Plaggen* of our farmers and those in Westphalia—that is, the outer crust of earth that one slices from the exposed sand of the heath, partly in order to improve the barren fields as much as possible, and partly for use in ovens. Therefore, the top of the heath is burnable, because its earth is infused with sprouting plants. The peat itself grew progressively through repetition of these *Plaggen*.

128. Cf. Plato, *Timaeus*, 48d–57c.

129. "Cheruscos": The Cherusci occupied the basin of the Weser River, north of the Chatti, in ancient times.

130. "Morinos": The Morini occupied parts of Flanders and Picardy.

[XLVII]

In Luneburgensi quoque agro, et alibi, nobis sub argilla latent arbores integrae vel fractae. Et memorabile est, fere uno situ jacere plerasque, radice inter Septentrionem et Occasum, cacumine inter Orientem et Meridiem porrectis. Idemque notavit Bootius, Brugensis, de patria sua: scilicet in fundis nonnullis dum ad decem aut etiam viginti ulnas foditur, integras reperiri sylvas, terra obrutas; agnosci exacte species arborum, et in foliorum serie annos distingui; truncos et folia pro carbonibus adhiberi, et arborum cacumina ad Orientem verti. Similia de Frisia memorantur et Groningano tractu. Itaque credunt viri docti, ante omnem annalium memoriam Oceanum aestu et Caecia Cauroque ventis furentem, quibus nunc quoque haec littora infestantur, magna vi irrupisse terris, unoque impetu totam hanc inferiorem Germaniam invecta materia obruisse, quam cum ex altiore loco venisse necesse sit, crediderim promontorium aliquod, aut naturales aggeres ex argilla mari objectos et tandem perfractos huc incubuisse.

[XLVIII]

Cum Amstelodami aliquando puteus foderetur ad ducentorum et triginta duorum pedum profunditatem, hae species terrarum ordine oblatae sunt: Hortensis terrae pedes septem, Torfae novem, argillae novem, arenae octo, terrae quatuor, argillae decem, terrae rursus quatuor, arenae, super qua domus illic fistucantur, pedes decem, argillae duo, sabulonis albi quatuor, siccae terrae quinque, turbidae unus, arenae quatuordecim, argillae arenariae tres, arenae cum argilla mistae quinque, arenae marinis conchyliis mistae quatuor. Tum fundus argillae ad centum et duorum pedum profunditatem; postremo triginta et unus sabulonis pedes; ubi fossio desiit. Ita Torfa semel occurrit, terra quinquies, argilla rursus quinquies, arena plus sexies, conchylia semel. Credibile est olim fundum maris fuisse, ubi nunc conchylia jacent ad centum amplius pedum profunditatem. Huic fundo reciprocatae inundationes, ruinaeque tot strata argillae arenaeque invexere, dum interim terrae sedimenta interjecti temporis

[XLVII. *On trees buried underground*]

In Lüneburg and elsewhere, every field has whole and broken trees hidden from us under clay. And it is remarkable that most of them lie in the same position, with the roots pointed between north and west, and the tips pointed between east and south. Bootius of Bruges[131] observed the same thing in his country: namely, in certain ground, after digging ten or even twenty ells down, one discovers whole forests covered by earth; the precise species of tree is recognized, and years are determined in the sequence of leaves; trunks and leaves are employed as coals, and the tips of the trees point east. The same is reported about Frisia and the province of Groningen. This is why learned men believe that in a time before all reported history, the boiling ocean, raging from the northeastern and northwestern winds that still attack these coasts today, burst onto the land with great force. And that one assault covered all of lower Germany with debris. Since this debris must have come from a higher place, I suppose that a promontory or a natural wall of clay blocked the sea; it was then broken apart and deposited here.

[XLVIII. *The layers of earth observed while digging a well in Amsterdam*]

When a well in Amsterdam was dug to a depth of two hundred and thirty-two feet, the kinds of earth were layered as follows: seven feet of garden earth, nine of peat, nine of clay, eight of sand, four of earth, ten of clay, another four of earth, ten feet of sand upon which the houses there are anchored, two of clay, four of white sand, five of dry earth, one of mud, fourteen of sand, three of sandy clay, five of sand mixed with clay, four of sand mixed with seashells. Then came one hundred and two feet of bottom clay, and finally thirty-one feet of coarse sand. The digging stopped there. Thus, peat occurred once, earth five times, clay also five times, sand more than six times, and shells once. In all likelihood, there was

131. Anselm Boetius de Boodt (ca. 1550–1634), a physician from Bruges who was also "gem advisor" to Rudolf II.

mora nascebantur. Sic repulsum mare cessit ad tempus, sed postea juris sui tenax, sese iterum ruptis aggeribus in terras infudit, sylvasque prostravit, quarum nunc ruinae a fodientibus deteguntur. Ita rerum natura praestat nobis Historiae vicem. Historia autem nostra hanc contra gratiam naturae rependit, ne praeclara ejus opera, quae nobis adhuc patent, posteris ignorentur.

once a seafloor where shells now lie, at a depth of more than one hundred feet. Repeated floods and catastrophes have thrown all the layers of clay and sand upon this floor, while the deposits of earth arose during the intervening periods. The sea, driven back, retreated for a time. But ultimately, insisting on its right, the sea once again burst the dams, flooding the lands and flattening the forests, whose ruins are now revealed by the diggers. For us, nature thus stands in place of history. But our written history repays nature's grace, so that her brilliant works, which still lie open before us, will not be ignored by posterity.

APPENDIX

Text from Friedrich Lachmund's

Oryktographia Hildesheimensis *(1669)*

Editors, starting with Scheidt (1749), inserted text from Lachmund directly into *Protogaea*. We include the relevant text from Lachmund here, as a separate appendix, since it does not appear in the A manuscript.[1] See the introduction.

1. Lachmund 1669, 40–56.

The strombite, or *Schneckenstein*, is similar to an aquatic snail. Indeed, it runs from wide to thin and ends in a spiral wound from the right. Sometimes it is short, sometimes nine inches long. (Kentmann calls the latter, in German, *Ein hoher und erhabner Schnecken-Stein;* the former he calls *ein zusammen gedruckter Schnecken-Stein*).[2] Inside it is white, and outside it takes on the color of the earth from which it was extracted. It is found in the quarries of Galgenberg, and in a new part of the city when digging the cellars where wine and beer are usually kept. Likewise, in Alfeld between the watchtower and the city, if one is turned toward Einbeck (Agricola, *De natura fossilium*).

The ctenite, *Kamstein*, or *Steinern Jacobs-Muscheln*, is grooved and looks just like a scallop. Its color is mostly ash gray. Found in the quarries beyond the Mount St. Moritz. (Agricola, loc. cit.). The Hildesheim ctenite, in the form of a whale mouth, is in Gessner, p. 165.[3]

The myite, or *Muschelstein*, looks like a mussel because it is not grooved. It is of two kinds, oblong and spherical, like a scallop. The latter, which is ash gray, is found in the quarries of our region, about which we have already spoken; the former, either brownish or yellowish, is extracted from the trench of the city's north-facing fortifications (Agricola, loc. cit.).

The onychite, in color and shape almost similar to the aromatic claws, which the Greeks call *onychas*, can be found in quarries (idem).

The ostreal stone, or *Ostren-Stein*, gets its name from oysters, which it resembles. It is of two kinds: a larger one that is easily split, like specular stone, and is dug out from the trench that runs north of the city, as I said; a smaller kind is found near the village of Linden, not far from Hannover (Agricola, loc. cit.). The larger kind can also be found in the quarries of Galgenberg and does not differ from real oysters.

The porphyroides, or *Purpur-Schnecken-Stein*, is spiked with barbs and is gray, like the purple sea snail. It is found in the city trench but is not wound like the purple sea snail. Another one is found there which is very similar to this one but does not have any barbs; but it has lateral stripes (idem).

The conchite, ornamented with curved ridges that converge to the

2. Johannes Kentmann (1518–1574). There are many grammatical and orthographic peculiarities in Lachmund's German terms. We have retained them here.

3. See Gessner 1565–1566.

back, and with gold-colored armature, is generally two palms long and one palm wide (idem).

I have found various stony shells here and there. Some of the more unusual ones from my collection are shown here.[4]

I. A gray ridged conchite: *eine halbe graue steinerne Muschel mit Strichen.*

II. Another the same color and a little smaller in size.

III. A gray, oblong myite: *ein halber langer Muschel-Stein.*

IV. A round conchite, yellowish, very small, and smooth: *ein klein gelber gantzer glatter Muschel-Stein.*

V. Another of the same color, a little larger.

VI. A round, gray, smooth myite, with another one embedded in it: *zwey halbe, runde Muschel-Steine ineinander.*

VII. A gray stone, with a round shell in it that is partly visible and partly hidden: *ein grauer Stein, darinn eine gantze runde Muschel.*

VIII. A gray and smooth conchite whose center is made of stone and can be removed and replaced again at will.

    a) The cochlite with its center.

    b) The same without its center.

    c) The stone center.

IX. A dark, grooved conchite with a thin shell. At the place where it is broken, it contains black soil; it is not much different from real shellfish.

X. A gray, smooth ostracite: *ein halber glatter Ostrenstein.*

XI. A yellowish rhomboid myite, similar to Rondelet's grooved mussel: *ein gantzer langer Muschel-Stein mit Strichen.*

XII. A round and smooth conchite: *ein runder gantzer glatter Muschel-Stein.*

XIII. A long and smooth yellowish conchite: *ein gantzer langer glatter Muschel-Stein.*

XIV. A smaller one, of the same color and shape.

XV. A small gray scallop with a thin shell, not very different from a real scallop: *ein klein Steinern Jacobs Muschel.*

XVI. A large, yellowish, wrinkled, conchite that has grooves going

4. This note from the Lachmund original does not appear in Ms XXIII, 23b, or in Leibniz 1749, nor is it included in any subsequent edition. The desciptions that follow apply to fig. 9 in *Protogaea*.

across from one edge to the other: *ein gantze runtzlichte runde steinern Muschel.*

XVII. A gray stone, which is entirely composed of round shells, either grooved or smooth; in it one can see the knotty horn of an ammonite. It is called by Jonston[5] "Megaric stone," because similar stones were once extracted in the Megara region, and used, as Pausanias tells it, to build a monument to Phoroneus and many buildings in the city of Megara (Agricola, *De natura fossilium*, book 7).

I. A long gray strombite made of hard, stony material, which is spread on roads: *ein langes graues steinernes Schneckenhaus.*

II. Another small one of the same color.

III. A yellow-brown one, not very hard, made of limestone.

IV. A yellow one, quite hard, made of limestone: *ein gelbhaffter langer Schneckenstein, als wenn er doppelt umgewunden wäre.*

V. A light red one, as smooth and elegant as if it had been turned on a lathe.

VI. Differs in size from IV [above].

VII. Is also made of limestone material.

VIII. A short strombite (cochlite) is a yellowish stone, similar to a land snail, composed of hard, calcareous material: *ein Stein den rechten Schneckenhaüsern gleich.*

IX. A dark snail stone, not so hard.

X. A small snail stone, the same color.

XI. A tube stone is a stone that looks exactly like the tubes of worms: *ein Wurmstein.*

XII. A dark brown cochlite, long and smooth, with worm tubes: *ein Steinern lange Muschel, mit Steinern Würmen.*

XIII. A dark brown snail with worm tubes: *ein Steinern Schneckenhaus, mit Steinern Würmen.*

A dark stone divided into two parts in its middle, one of which shows an elegantly grooved snail, while the other shows its matrix, which, when joined to the snail, fits it like a mold. It was found in a gravel quarry. *Ein brauner Stein, in der mitte voneinander gespalten, in welches einem Theil sitzet eine schöne Krause Schnecke im andern derselben Form, kan voneinander, und wieder zusammengeleget werden.*

A dark brown, stony mass with molds of grooved snails everywhere,

---

5. Jan Jonston (1603–1675), friend and patron of Comenius.

partly hidden and partly visible. They look so elegant that they could not have been better shaped by the hand of the most skillful sculptor.

The trochite, *Spangen-, Räder-, Zwerg-, oder Mühlstein* resembles the Jewstone (as Agricola says) and is called that because it looks like a wheel. And indeed, nature has given it the shape of a wheel: its round part is smooth, and from the center of its cross section spokes reach toward the outside of the disk, just like those of a wheel, and they jut out so much that grooves are created. It varies greatly in size, with the biggest being ten times bigger than the smallest: the biggest is as broad as a finger and as thick as the third part of it, or a little more. It can be of different colors, either gray, black, or clay colored, but they usually get that way because of contact with the earth; inside they are whiter than the others. All their fractures are smooth and shiny, as with the Jewstone, and they break in the same way in their length, width, or diagonally; if placed in vinegar, they produce bubbles, like astroite. One even occasionally finds some of them that move from their place, like astroites. Entrochites are made of trochites that are not yet separated (*Spangenstein aneinander*), numbering three, sometimes four, or even more. Twenty attached in a group have even been found. They are of two kinds: either completely cylindrical, or somewhat cylindrical, with their middle partly swollen and both ends thinner. In those with larger spokes, there always seems to be a twisted belt where the two are connected. Those with smaller spokes lack the belt and are completely smooth. Trochites are joined together so that the spokes of the one meet the grooves of the other. Trochites of swollen entrochites generally have quite small spokes. It is not unusual to find, together with a trochite and an entrochite, a shapeless stone that contains the figure of a wheel, remaining as a sort of impression, the trochite having broken away. Saxony produces those stones at Hildesheim, beyond Moritzberg, in the cracks of gray-white marble and in clay soil. The same can be found between Alfeld and Einbeck. Trochites break up kidney stones and are useful for curing urinary problems.

I have found them in great abundance, and I show the rarest of them here.[6]

6. Lachmund's key here corresponds to fig. 11 in *Protogaea*.

I.   1. A small, very pale yellow trochite.
     2. A larger one, the same color.
     3. A gray one in the shape of a wheel.
     4. Another swollen in the middle.
     5. Another in the shape of a rose.
     6. A large, white one with protruding spokes.
     7. Similar to the last one in size, but twice as thick.
     8. A small one shaped like a column (Ferrante Imperato, in Worm. Mus., p. 70, columnetta).
     9. Another, different from the last in thickness and length. This may be the small Jewstone that is shaped like a cylinder and mentioned in Reeland, Lex. Alchem., p. 283. *Er siehet wie ein klein Mühlstein.*
     10. Another in the shape of a penis, without its prepuce.
II. 1. A small entrochite composed of two trochites.
     2. One trochite joined to another along its side.
     3. Another of moderate size.
     4. Another blackish one, hollow in the middle but swollen at the ends.
III. 1. An entrochite composed of three trochites, without the twisted girdle.
     2 and 3 differ only in size.
IV. 1. Consists of four (trochites).
     2. The same, a little larger. The first is not in the shape of a wheel, but starts with a wide base and ends in a sharp point.
     3. Composed of very large and whitish trochites.
V.   1. Composed of five trochites, the first of which lacks the twisted girdle.
     2 and 3 differ only by their larger size.
VI. 1. Composed of six white trochites, with a swollen middle section.
     2. Gray, with a point standing out on the the first (trochite).
VII. Composed of eight, the first of which is broken and has a curved base.
VIII. Made of eight white trochites, some of which are thicker and are alternatingly joined to the others.
IX. Composed of five, but the two upper ones are nested differently from the others.

x. A gray stone with a ctenite joined to a white, striated trochite.

xi. A smooth, gray, round conchite; it has a circle with a spot in its center and is the base of a broken trochite.

xii. A gray stone, containing several trochites and entrochites.

GLOSSARY

Like all words, these terms have discrete historical meanings that cannot be reduced to modern or "correct" definitions. Historical definitions or equivalents thus appear first. Modern definitions, where appropriate, appear after these, in parentheses.

acetum. Vinegar.

aes. Copper or copper ore.

alabastrita. Alabaster, a fine-grained sulphate of lime or gypsum.

alumen. Alum. (A mineral salt, the double sulfate of aluminum and potassium. Used for dyeing and medicinal purposes.)

amianthus. A variety of asbestos, characterized by long, flexible, pearly-white fibers.

ambra. Amber. See *succinum.*

ammoniacum. Sal ammoniac or the salt of Ammon. (A salt composed mostly of ammonium chloride.)

antimonium. The native ore of antimony, or stibnite.

ardesia, ardosia. Copper schist.

argentum capillare. Natural silver with a hairlike or capillary appearance.

argentum rude rubrum. Raw red silver.

argentum vitriforme. Argentite. A glasslike silver ore.

argentum vivum. Quicksilver, or mercury.

asterias. Asterias, a star-shaped petrification. (A kind of urchin, or echinoderm, like the common starfish. A starlike fragment of the stem of a fossil crinoid.)

astroites. An astroite. (A coral stone, or any star-shaped mineral or fossil.)

auripigmentum. Orpiment, auripigment, yellow arsenic. (It occurs naturally in soft, gold-colored masses. Orpiment was used mainly as a pigment.)

belemnites. Belemnite, a petrification of oblong and cylindrical shape. (The endoskeleton of an extinct squidlike cephalopod). See *lyncurium.*

bitumen. Bitumen, asphalt, or mineral pitch.

brontia. A "toadstone." (A spherical and hollow stone.)

bryonia. Bryony, a vine used for medicinal purposes.

buccina. A shell shaped like a tromb. (The twisted and spiral shell of a mollusk. A whelk.)

butter of antimony. Prepared by distilling antimony and a sublimate (mercuric chloride). (In modern chemical terms, antimony trichloride.)

cali. Kali, soda ash, vegetable alkali, or potash.

canis marinus. A sea dog. (A shark.)

carbunculus. A carbuncle, that is, a ruby or red precious stone.

chalcantum. Native blue vitriol or sulfate of copper.

cinnabaris. Cinnabar, a bright red stone which is an ore of mercury. It is the red form of mercuric sulfide.

cinnabaris antimonialis. Cinnabar of antimony, formed in the process of making butter of antimony.

cobaltum, coboltum. Cobalt ore, especially as it was found naturally in the veins of the Erz Mountains.

cobalti fumus. The fume of cobalt. (Arsenic that is contained in the ore cobaltite.)

cochlites. A snail-like petrification. (A fossil spiral shell or sea snail.)

conchites. Petrification in the form of a trumpet. (A mollusk or brachiopod.)

cornu ammonis. Ammon's horn or snakestone. (Ammonite, an extinct genus of cephalopod, with a shell curved like a ram's horn.)

ctenites. A comblike petrification. (An animal with ctenoid features, like a scallop or pecten.)

echinus. Sea urchin.

empyreuma. The burnt-smelling materials produced during a distillation.

entale. A kind of snail.

entrochus. Entrochite. (Part of the stem of a fossil crinoid, or sea lily.)

fluor caeruleus. Blue fluorspar.

gagates. Jet. (A black form of lignite.)

galena. Native lead sulfide, a common lead ore.

glossopetrae. Glossopetrae, petrified snakes' tongues. (Fossilized sharks' teeth.)

haematites. Hematite, or bloodstone. (An abundant iron ore naturally occurring in various colors, especially red and reddish brown.)

histrix. The fossil shell of a brachiopod.

hydrargyrum. Mercury or quicksilver.

lapis calaminaris. Calamine, an ore of zinc.

lapis judaicus. A Jewstone, a petrification found near Palestine. (The
   fossil spine of a large sea urchin.)

lapis tophaceus. Tuff stone, a light and porous cellular rock.

limus. Clay, clayey earth, mud.

lithantracum. Hard coal.

lixivium. Lye or potash. The liquid product obtained by leaching the
   ashes of vegetables and evaporating the solution.

lyncurium. Lynx stone, a yellowish stone supposed to have originated in
   lynx urine. See *belemnites.*

mandragora. Mandragora, or mandrake.

marga. Clay or marl.

melanteria. Melanterite or iron vitriol. (Ferrous iron sulfate.)

mercurius. Mercury, or quicksilver.

milleporum. A millepore. (A coral with a large calcareous skeleton.)

myites. A mussel-like petrification. (Brachiopods, shells.)

naptha. Naptha or liquid petroleum.

nitrum. Saltpeter or niter. See *nitri spiritus.*

nitri spiritus. Spirit of niter, also known as spirit of saltpeter. Niter,
   or "volatile niter," was also sometimes regarded as a component
   of the atmosphere that supplied the world with a principle of life.
   (Nitric acid.)

onychites. Onyx. (A variety of marble or alabaster, with yellow veins.)

ossifraga. Osteocolla, called "glue-bone stone," because of its supposed
   ability to knit broken bones. (A calcareous agglomeration of roots
   and stems. Often found in sandy ground.)

osteocolla. See *ossifraga.*

ostracis lapis. A petrification in the shape of an oyster. (A fossilized
   oyster.)

per deliquium. Refers to the dissolution of a solid body through humidity,
   often in a cold, wet place, like a cellar.

phosphorus smaragdinus. Green phosphorus, that is, fluorspar that glows
   when warmed.

plumbum nigrum. White lead ore, or cerussite, darkened by galena.

porphyroides. Purple sea snail. (Murex, a marine gastropod.)

pyrita, pyrites. Pyrites, or firestones. (Sulfides of iron and copper.)

regulus. The metallic component of an ore.

rubrica fabrilis. Red chalk or red ocher. A fine red or reddish brown clay.

saccarus saturni. The sugar of lead.

sal de cornu cervi. The salt of hartshorn. (Ammonium carbonate.)

sal gemmeum. Rock salt.

sandaraca rubra. Red sandarac or realgar.

saturnus. Generally lead, but could also refer to stibnite.

schistus. Schist, slate, or hornblende.

specularis lapis. Specular stone. (A species of mica, selenite, or talc.)

stenomarga. A light-colored clay.

stiria. A concretion, such as a stalactite, that resembles an icicle.

strombites. Strombites. (Fossilized stromb shells.)

succinum. Yellow amber. (A fossil resin of conifer trees.)

terra tartari foliata. Regenerated tartar. (Potassium acetate.)

tophus. Tuff stone.

torfa, turfa. Peat.

trochites, trochita. Trochites, or wheel-stones. (The wheel-like joints of encrinites, that is, of fossil crinoids.) See *entrochus.*

trochoides. Trochoids or conical shells.

tubulites. Petrifications with tubes. (Fossil tubular shells, the tubular shells of the shipworms.)

turbo. A mollusk with a whorled shell.

vitriolum. Vitriol. (Copper sulfate.)

zafera. Zaffer. (An impure oxide of cobalt.)

BIBLIOGRAPHY

ARCHIVAL DOCUMENTS

Gottfried Wilhelm Leibniz Bibliothek, Hannover, Handschriftenbestand, Ms XXIII, 23a and 23b. Includes the "A manuscript" (copy with revisions in Leibniz's own hand) and Eckhart's revised version.

Forschungsbibliothek Gotha, Chart B 199, Papers of Wilhelm Ernst Tenzel. This fascicle includes letters from G. W. Leibniz about the history of the earth and fossil objects.

Oberbergamt Clausthal-Zellerfeld, Fach 761, Acta 27, "Leibniz's Windmühlenkünste"; and Fach 762, Acta 29, "Leibniz's Windmühlenkünste."

PRIMARY SOURCES

Agricola, Georgius. 1546. *De ortu et causis subterraneorum . . . De natura fossilium.* Basel.

———. [1546] 1955. *De natura fossilium.* Translated by Mark Chance Bandy and Jean A. Bandy. New York.

———. [1556] 1912. *De re metallica.* Translated by Herbert and Lou Henry Hoover. London.

Albinus, Petrus. 1590. *Meißnische Land- und Berg-Chronica.* Dresden.

Bartholin, Thomas. 1645. *De unicornu observationes novae.* Poitiers.

Becher, Johann J. 1681. *Actorum laboratorii chymici Monacensis seu Physicae subterraneae libri duo.* Frankfurt.

Bernier, François. [1684] 1992. *Abrégé de la philosophie de Gassendi.* 7 vols. Paris.

Boyle, Robert. 1661. *The Sceptical Chymist.* London.

———. 1669. *Certain Physiological Essays and Other Tracts.* London.

Buffon, Georges Louis Leclerc de. 1779. *Histoire naturelle.* Paris.

Burnet, Thomas. 1681. *Telluris theoria sacra orbis nostri originem & mutationes generales, quas aut jam subiit, aut olim subiturus est, complecten: Libri duo priores de diluvio & paradiso.* London.

Calvör, Henning. 1763. *Acta historico-chronologico-mechanica circa metallurgicam in Hercynia Superiori, oder, Historisch-chronologische Nachricht und theoretische und practische Beschreibung des Maschinenwesens, und der Hülfsmittel bey dem Bergbau auf dem Oberharze.* Braunschweig.

Cavalieri, Bonaventura. 1632. *Lo specchio vstorio; overo, Trattato delle settioni coniche, et alcvni loro mirabili effetti intorno al lume, caldo, freddo, suono, e moto ancora.* Bologna.

Chambers, Ephraim. 1728. *Cyclopaedia, or, an Universal Dictionary of Arts and Sciences.* 2 vols. London.

Descartes, René. [1633] 1979. *Le Monde.* Translated by Michael Sean Mahoney. New York.

———. [1644] 1983. *Principles of Philosophy.* Translated by Valentine Rodger Miller and Reese P. Miller. Dordrecht.

Fénelon, François de Salignac de La Mothe. [1710s] 1897. *Lettre sur les occupations de l'Académie française.* Paris.

Fontenelle, Bernard Le Bovier de. [1735] 1766. "Éloge de Bianchini." In *Éloges des académiciens de l'Académie royale des sciences.* Paris.

Gassendi, Pierre. 1641. *Viri illustris Nicolai Claudii Fabricii de Peiresc.* Paris.

Gessner, Konrad. 1565–1566. *De omni rerum fossilium genere, gemmis, lapidibus, metallis, et huiusmodi, libri aliquot, plerique nunc primum editi.* Zurich.

González de Salas, Jusepe Antonio. 1650. *De duplici viventium terra disputatio paradoxica.* Leiden.

Guericke, Otto von. 1672. *Experimenta nova (ut vocantur) Magdeburgica de vacuo spatio.* Amsterdam.

Kant, Immanuel. [1786] 2004. *The Metaphysical Foundations of Natural Science.* Translated and edited by Michael Friedman. Cambridge.

Kircher, Athanasius. 1664. *Mundus subterraneus.* Amsterdam.

Lachmund, Friedrich. 1669. *Oryktographia Hildesheimensis.* Hildesheim.

Leeuwenhoek, Antony van. 1677–1678. "Observationes D. Anthonii Lewenhoeck de natis e semine genitali animaculis." *Philosophical Transactions of the Royal Society* 12, no. 142:1040–1046.

Leibniz, Gottfried Wilhelm von. 1693. "Protogaea." *Acta eruditorum,* pp. 40–42.

———. 1710. *Miscellanea Berolinensia* 1:118–120.

———. [1710] 1966. *Essais de Theodicée sur la bonté de Dieu, la liberté de l'homme et l'origine du mal.* Amsterdam. English translation, *Theodicy,* translated by E. M. Huggard, edited by Diogenes Allen (Indianapolis, 1966).

———. 1720. *Lehrsätze über die Monadologie.* Jena.

———. 1749. *Protogaea; sive, De prima facie telluris et antiquissimae historiae vestigiis in ipsis naturae monumentis dissertatio.* Edited by Christian Ludwig Scheidt. Göttingen. French translation, *Protogée ou De la formation et des révolutions du globe,* edited by Bertrand de Saint-Germain (Paris, 1859). German translation, *Protogaea,* edited and translated by W. von Engelhardt (Stuttgart, 1949). French translation, *Protogaea: De l'aspect primitif de la terre et des traces d'une histoire très ancienne que renferment les monuments mêmes de la nature,* edited by Jean-Marie Barrande (Toulouse, 1993).

———. 1768. *Leibnitii opera omnia.* Edited by L. Dutens. 6 vols. Geneva.

———. 1923–. *Gottfried Wilhelm Leibniz: Sämtliche Schriften und Briefe.* Darmstadt, Deutsche Akademie der Wissenschaften.

———. 1960. *Die philosophischen Schriften.* Edited by C. I. Gerhardt. 7 vols. Berlin, 1875–1890; reprint, Hildesheim.

———. 1971. *Briefwechsel zwischen Leibniz und Christian Wolff.* Edited by C. I. Gerhardt. Hildesheim.

———. 1989. *Philosophical Essays.* Edited and translated by Roger Ariew and Daniel Garber. Indianapolis.

———. 2000. *G. W. Leibniz and Samuel Clarke: Correspondence.* Edited by Roger Ariew. Indianapolis.

Leupold, Jacob. 1725. *Theatrum machinarum generale.* Leipzig.

Maillet, Benoît de. [1748] 1755. *Telliamed.* Amsterdam.

Marana, Giovanni Paolo. 1710. *L'espion dans les cours des princes chrétiens.* 6 vols. Cologne.

Mersenne, Marin. 1634. *Questions inouyes; ou, Récréation des scavans.* Paris.

Montfaucon, Bernard de. 1719. *L'Antiquité expliquée et representée en figures.* 5 vols. Paris.

Nibelungenlied. 1999. Translated by Margaret Armour. Cambridge, Ont.

Ramazzini, Bernardino. 1691. *De fontium Mutinensium admiranda scaturigine.* Modena.

Ray, John. 1692. *Miscellaneous Discourses Concerning the Dissolution and Changes of the World.* London.

Scheidt, Christian Ludwig, ed. 1750–1753. *Origines Guelficae.* Göttingen.

Schönberg, Abraham von. 1693. *Ausführliche Berg-Information.* Leipzig.

Schröder, Wilhelm Freiherr von. 1713. *Fürstliche Schatz- und Rent-Kammer.* Leipzig.

Scilla, Agostino. 1670. *Vana speculazione disingannata dal senso.* Naples.

Seckendorff, Veit Ludwig von. [1665] 1976. *Teutscher Fürsten-Stat.* 2 vols. Glashütten.

Steno, Nicolaus. [1667] 1969. "A Carcharadon-Head Dissected." In *Geological Papers,* edited by Gustav Scherz, translated by Alex J. Pollock, 72–117. Odense.

———. 1669. *De Solido intra solidum naturaliter contento dissertationis prodromus.* Florence.

———. [1669] 1916. *The Prodromus of Nicolaus Steno's Dissertation concerning a solid body enclosed by process of nature within a solid.* Edited by J. G. Winter. London.

Tavernier, Jean-Baptiste. 1678. *Les six voyages en Turquie, en Perse et aux Indes.* Paris.

Trebra, Friedrich Wilhelm Heinrich von. [1818] 1990. *Bergmeister-Leben und Wirken in Marienberg.* Leipzig.

Tschirnhaus, E. F. 1697. "De magnis lentibus seu vitris causticis." *Acta eruditorum,* pp. 414–419.

Zedler, Johann Heinrich, ed. 1732–1750. *Grosses vollständiges Universal-Lexicon aller Wissenschafften und Kunste.* 64 vols. Halle.

Zimmermann, C. F. 1746. *Ober-Sächsische Berg-Akademie.* Dresden.

SECONDARY LITERATURE

Abel, Othenio. 1925. *Geschichte und Methode der Rekonstruktion vorzeitlicher Wirbeltiere.* Jena.

Adams, Frank Dawson. 1938. *The Birth and Development of the Geological Sciences.* Baltimore.

Aiton, E. J. 1985. *Leibniz: A Biography.* Bristol.

Ariew, Roger. 1991. "A New Science of Geology in the Seventeenth Century?" In *Revolution and Continuity: Essays in the History and Philosophy of Early Modern Science,* edited by Peter Barker and Roger Ariew, 81–92. Washington, D.C.

———. 1988. "Leibniz's Protogaea." In *Leibniz: Tradition und Aktualität: V. Internationaler Leibniz-Kongress,* edited by Ingrid Marchlewitz. Hannover.

———. 1998. "Leibniz on the Unicorn and Various Other Curiosities." *Early Science and Medicine* 3:267–288.

Barret-Kriegel, Blandine. 1988. *La Défaite de l'érudition.* Paris.

Bartels, Christoph. 1992. *Vom frühneuzeitlichen Montangewerbe zur Bergbauindustrie: Erzbergbau im Oberharz, 1635–1866.* Bochum.

———. 1997. "Die Nutzung der Wasserkraft im Harzer Montanwesen." In *Naturwissenschaft und Technik im Barock,* edited by Uta Lindgren, 51–76. Cologne.

Baumgärtel, Hans. 1965. "Vom Bergbüchlein zur Bergakademie: Zur Entstehung der Bergbauwissenschaften zwischen 1500 und 1765/1770." *Freiberger Forschungshefte* D50.

Belaval, Yvon. 1975. *Leibnitz: Initiation à sa philosophie.* Paris.

Bozorgnia, S. M. H. 1998. "The Role of Precious Metals in European Economic Development." *Contributions in Economics and Economic History* 192:164–165.

Breger, Herbert, and Friedrich Niewöhner, eds. 1999. *Leibniz und Niedersachsen.* Stuttgart.

Burose, Hans. 1967. "Markscheider Bernhard Ripking: Sein Leben, sein Wirken und sein Briefwechsel mit G. W. v. Leibniz." *Der Anschnitt* 19, no. 5:17–25.

Cavaillé, Jean-Pierre. 1991. *Descartes: La Fable du monde.* Paris.

Cohen, Claudine. 1996. "Leibniz Protogaea: Patronage, Mining and the Evidence

for a History of the Earth." In *Proof and Persuasion*, edited by Suzanne
Marchand and Elizabeth Lundbeck, 125–143. Amsterdam.

———. 1997. "Des fossiles et des mines." *La Revue du Centre National des Arts et
Métiers* 18 (March 1997): 4–16.

———. 1998. "An Unpublished Manuscript by Leibniz (1646–1716) on the Nature
of 'Fossil Objects.'" *Bulletin de la Société Géologique de France* 169, no. 1:137–142.

———. 2002. *The Fate of the Mammoth*. Chicago.

———. 2008. *La Genèse de Telliamed: Science, libertinage et clandestinité à l'aube des
Lumières*. Paris.

Conze, Werner. 1951. *Leibniz als Historiker*. Berlin.

Cooper, Alix. 2003. "'The Possibilities of the Land': The Inventory of 'Natural
Riches' in the Early Modern German Territories." *History of Political Economy*
35, annual suppl.: 129–153.

———. 2007. *Inventing the Indigenous: Local Knowledge and Natural History in Early
Modern Europe*. Cambridge.

Davillé, Louis. [1909] 1986. *Leibniz historien: Essai sur l'activité et la méthode
historiques de Leibniz*. Aalen.

Duhem, Pierre. 1913–1959. *Le système du monde: Histoire des doctrines
cosmologiques de Platon à Copernic*. 10 vols. Paris.

Ellenberger, François. 1988–. *Histoire de la géologie*. Paris.

Elster, Jon. 1975. *Leibniz et la formation de l'esprit capitaliste*. Paris.

Faak, Margot. 1980. "Leibniz' Bemühungen um die Reichshofratswürde in den
Jahren 1700 bis 1701." *Studia Leibnitiana* 12:114–124.

Forberger, Rudolf. 1964. "Johann Daniel Crafft: Notizen zu einer Biographie."
Jahrbuch für Wirtschaftsgeschichte 2/3:63–79.

Garber, Daniel. 1985. "Leibniz and the Foundations of Physics: The Middle
Years." In *The Natural Philosophy of Leibniz*, edited by Kathleen Okruhlik and
James Robert Brown, 27–130. Dordrecht.

———. 1988. "Force and the Relativity of Motion in Leibniz's Physics."
In *Leibniz: Tradition und Aktualität: V. Internationaler Leibniz-Kongress*.
Hannover.

———. 1995. "Leibniz: Physics and Philosophy." In *The Cambridge Companion to
Leibniz*, edited by Nicholas Jolley, 270–352. Cambridge.

———. 1997. "Leibniz on Form and Matter." *Early Science and Medicine* 2:
326–352.

Garber, Daniel, and Roger Ariew. 1992. "Descartes' Physics." In *The Cambridge
Companion to Descartes*, edited by J. Cottingham, 286–334. Cambridge.

———, eds. 1998. "Leibniz and the Sciences." *Perspectives on Science* 6, nos. 1 and 2.

Garber, Daniel, and Michael Ayers, eds. 1998. *The Cambridge History of
Seventeenth-Century Philosophy*. Cambridge.

Gohau, Gabriel. 1991. *A History of Geology.* Revised and translated by Albert V. Carozzi and Marguerite Carozzi. New Brunswick, N.J.

Gottschalk, Jürgen. 1977. "Bemerkungen zu Hans Peter Münzenmeyers Beitrag "Leibniz' Inventum Memorabile: Die Konzeption einer Drehzahlregelung vom März 1686." *Studia Leibnitiana* 9:275–278.

———. 1999. "Der Oberharzer Bergbau und Leibniz' Tätigkeit für Verbesserungen." In *Leibniz und Niedersachsen,* edited by Herbert Breger and Friedrich Niewöhner, 173–186. Stuttgart.

———. 2000. "Technische Verbesserungsvorschläge im Oberharzer Bergbau." In *Gottfried Wilhelm Leibniz: Das Wirken des großen Universalgelehrten als Philosoph, Mathematiker, Physiker, Techniker,* edited by K. Popp and E. Stein, 109–132. Hannover.

Gould, Stephen J. 1987. *Time's Arrow and Time's Cycle: Myth and Metaphor in the Discovery of Geological Time.* Cambridge, Mass.

———. 2002. *I Have Landed: The End of a Beginning in Natural History.* New York.

Grene, Marjorie. 1985. *Descartes.* Brighton, England.

Hamm, Ernst P. 1993. "Bureaucratic 'Statistik' or Actualism? K.E.A. von Hoff's 'History' and the History of Geology." *History of Science* 31:151–176.

———. 1997. "Knowledge from Underground: Leibniz Mines the Enlightenment." *Earth Sciences History* 16:77–99.

———. 2001. "Unpacking Goethe's Collections: The Public and the Private in Natural-Historical Collecting." *British Journal for the History of Science* 34:275–300.

Hermann, Walther. 1962. "Bergrat Henckel. Ein Wegbereiter der Bergakademie." *Freiberger Forschungshefte* D37.

Horst, Ulrich, and Jürgen Gottschalk. 1973. "Über die Leibniz'schen Pläne zum Einsatz seiner Horizontalwindkunst im Oberharzer Bergbau." In *Akten des II. Internationalen Leibniz-Kongresses,* edited by Albert Heinekamp, Dorothea Kalisch, et al., 35–69. Wiesbaden.

Huxley, Thomas Henry. 1880. "On the Method of Zadig." In *Collected Essays,* vol. 4.

Jolley, Nicholas, ed. 1995. *The Cambridge Companion to Leibniz.* Cambridge.

Knobloch, Eberhard. 1997. "Leibniz als Wissenschaftspolitiker: Vom Kulturideal zur Societät der Wissenschaften." In *Naturwissenschaft und Technik im Barock,* edited by Uta Lindgren, 99–112. Cologne.

Kraschewski, Hans-Joachim. 1995. "Das Direktionsprinzip im Harzrevier des 17. Jahrhunderts und seine wirtschaftspolitische Bedeutung." In *Vom Bergbau- zum Industrierevier,* edited by Ekkehard Westermann, 125–150. Stuttgart.

Laming-Emperaire, Annette. 1965. *Origines de l'archéologie préhistorique en France.* Paris.

Laudan, Rachel. 1987. *From Mineralogy to Geology*. Chicago.

Lindgren, Uta, ed. 1997. *Naturwissenschaft und Technik im Barock*. Cologne.

Lommatzsch, Herbert. 1968. "Gottfried Wilhelm Leibniz als Erfinder im Harz." In *Erfindungen im harzer Erzbergbau*, edited by A. Reichers, 15–24. Clausthal-Zellerfeld.

Meyer, R. W. 1952. *Leibnitz and the Seventeenth-Century Revolution*. Translated by J. P. Stern. Cambridge.

Morello, Nicoletta. 1979. *La nascita della paleontologia nel seicento: Colonna, Stenone e Scilla*. Milan.

Münzenmeyer, Hans Peter. 1976. "Leibniz' Inventum Memorabile: Die Konzeption einer Drehzahlregelung vom März 1686." *Studia Leibnitiana* 8, no. 1:113–119.

Nef, John U. 1941. "Silver Production in Central Europe, 1450–1618." *Journal of Political Economy* 49:575–591.

Newman, William R., and Lawrence M. Principe. 1998. "Alchemy versus Chemistry: The Etymological Origins of a Historiographic Mistake." *Early Science and Medicine* 3:32–65.

———. 2002. *Alchemy Tried in the Fire: Starkey, Boyle, and the Fate of Helmontian Chymistry*. Chicago.

———, eds. 2004. "Introduction." In *Alchemical Laboratory Notebooks and Correspondence*, by George Starkey, edited by William R. Newman and Lawrence M. Principe, ix–xxix. Chicago.

Ogilvie, Brian W. 2006. *The Science of Describing: Natural History in Renaissance Europe*. Chicago.

Oldroyd, David. 1996. *Thinking about the Earth: A History of Ideas about Geology*. London.

Oldroyd, David, and J. B. Howes. 1978. "The First Published Version of Leibniz's 'Protogaea.'" *Journal of the Society for the Bibliography of Natural History* 9:56–90.

Rappaport, Rhoda. 1991. "Fontenelle Interprets the Earth's History." *Revue d'Histoire des Sciences* 44:281–300.

———. 1997a. "Leibniz on Geology: A Newly Discovered Text." *Studia Leibnitiana* 29, no. 1:6–11.

———. 1997b. *When Geologists Were Historians*. Ithaca, N.Y.

Rescher, Nicholas. 1999. "Leibniz Visits Vienna (1712–1714)." *Studia Leibnitiana* 31, no. 2:133–159.

Ritter, Paul. 1938. "Einleitung." In G.W. Leibniz: *Sämtliche Schriften und Briefe*, ed. Preussische Akademie der Wissenschaften, ser. 1, vol. 3, xxvii–xlvi. Leipzig.

Robinet, André. 1988. *G. W. Leibniz: Iter italicum (mars 1689–mars 1690): La Dynamique de la République des lettres: Nombreux Textes inédits*. Florence.)

Roger, Jacques. 1968. "Leibnitz et la Théorie de la Terre." In *Leibnitz, 1646–1716: Aspects de l'homme et de l'oeuvre*, 137–144. Paris.

Ross, George MacDonald. 1974. "Leibniz and the Nuremberg Alchemical Society." *Studia Leibnitiana* 6:222–248.

———. 1984. *Leibniz*. Oxford.

Rossi, Paolo. 1984. *The Dark Abyss of Time*. Chicago.

Rudwick, Martin J. R. 1972. *The Meaning of Fossils*. Chicago.

Russel, Bertrand. 1900. *A Critical Exposition of the Philosophy of Leibniz*. London.

Rutherford, Donald. 1995. "Metaphysics: The Late Period." In *The Cambridge Companion to Leibniz*, edited by Nicholas Jolley, 143–153. Cambridge.

Scheel, Günter. 1976. "Leibniz und die deutsche Geschichtswissenschaft um 1700." In *Historische Forschung im 18. Jahrhundert*, edited by Karl Hammer and Jürgen Voss, 82–101. Bonn.

———. 1991. "Einleitung." In *Sämtliche Schriften und Briefe: Supplementband: Harzbergbau 1692–1696*, by Gottfried Wilhelm Leibniz, ser. 1, xxvii–lxv. Berlin.

Schnapp, Alain. 1993. *La conquête du passé: Aux origines de l'archéologie*. Paris.

Schnapper, Antoine. 1988. *Le géant, la licorne, la tulipe: Collections et collectionneurs français du XVIIème siècle*. Paris.

Skeat, Walter W. 1912. "'Snakestones' and Stone Thunderbolts as Subjects for Systematic Investigation." *Folklore* 23, no. 1:45–80.

Smith, Pamela H. 1994a. *The Business of Alchemy*. Princeton, N.J.

———. 1994b. "Alchemy as a Language of Mediation at the Habsburg Court." *Isis* 85, no. 1:1–25.

Soetbeer, Adolf. 1879. *Edelmetall-Produktion und Wertverhältniss zwischen Gold und Silber seit der Entdeckung Amerikas bis zur Gegenwart*. Gotha.

Solinas, Giovanni. 1973. "La Protogaea di Leibnitz ai margini della rivoluzione scientifica." In *Saggi sull'illuminismo*, edited by Giovanni Solinas, 7–70. Cagliari.

Spitz, Lewis W. 1952. "The Significance of Leibniz for Historiography." *Journal of the History of Ideas* 13, no. 3:333–348.

Sticker, Bernhard. 1967. "Leibniz' Beitrag zur Theorie der Erde." *Sudhoffs Archiv* 51, no. 3:244–259.

Stiegler, Leonhard. 1968. "Leibnizens Versuche mit der Horizontalwindkunst auf dem Harz." *Technikgeschichte* 35, no. 4:265–292.

Troitzsch, Ulrich. 1966. *Ansätze technologischen Denkens bei den Kameralisten des 17. und 18. Jahrhunderts*. Berlin.

Wakefield, Andre. 2000. "Police Chemistry." *Science in Context* 13, no. 2:231–267.

———. 2002. "Abraham Gottlob Werner and the Cameralist Tradition in Freiberg." *Freiberger Forschungshefte* D207:379–387.

————. 2007. "The Fiscal Logic of Enlightened German Science." In *Making Knowledge in Early Modern Europe*, edited by Benjamin Schmidt and Pamela Smith. Chicago.

Weber, Wolfhard. 1976. *Innovationen im Frühindustriellen deutschen Bergbau und Hüttenwesen: Friedrich Anton von Heynitz*. Göttingen.

Yamada, Toshihiro. 2001. "Leibniz's Unpublished Drawings in Protogaea Manuscript." *JAHIGEO Newsletter* 3:4–6.

Zittel, Karl Alfred von. 1901. *History of Geology and Palaeontology to the End of the Nineteenth Century*. London.

INDEX

Page numbers in italics denote illustrations.